Counter Electrode for Dye-Sensitized Solar Cells

Counter Electrode for Dye-Sensitized Solar Cells

edited by

Alagarsamy Pandikumar

Ramesh Mohan

Kandasamy Jothivnekatachalam

JENNY STANFORD
PUBLISHING

Published by

Jenny Stanford Publishing Pte. Ltd.
Level 34, Centennial Tower
3 Temasek Avenue
Singapore 039190

Email: editorial@jennystanford.com
Web: www.jennystanford.com

British Library Cataloguing-in-Publication Data
A catalogue record for this book is available from the British Library.

Counter Electrode for Dye-Sensitized Solar Cells

ISBN 978-981-4877-38-1 (Hardcover)
ISBN 978-1-003-11077-4 (eBook)

Contents

Preface

Energy is one of the most important challenges that humanity is confronted with today. The global worry over this energy problem can mainly be ascribed to the accelerating economic growth, ecological factors, and exhaustible fossil fuels. Unruffled economic growth at a high speed requires more energy support: exhaustible fossil fuels are being widely used now and the demand has moved toward renewable energy sources. The research and development in the field of renewable energy resources have received great attention worldwide. The sun is the leader among all renewable energy sources, and the upper atmosphere of the earth receives 174 petawatts (PW) of solar radiation in one year. This energy amounts to twice that obtained from non-renewable resources of coal, oil, natural gas, and uranium. Solar photovoltaics turned into a fast-growing industry and the most potential of any other renewable technologies.

Three generations of photovoltaic technology (solar cells) have evolved so far; the first-generation cells are conventional cells made of monocrystalline or polycrystalline silicon solar cell whereas the second-generation solar cells involve thin-film technology such as copper indium gallium selenide and cadmium telluride. The emerging photovoltaic technology describes the third-generation solar cells as those including dye-sensitized solar cells (DSSCs), perovskite solar cells, organic/polymer solar cells, and quantum solar cells. These third-generation solar cells have low environmental impact as well as processing cost. DSSCs are one of the most promising third-generation solar cell techniques to harvest solar energy and convert into electrical energy. DSSC consists of four main components: (i) photoanode, (ii) sensitized dye, (iii) redox couple electrolyte, and (iv) counter electrode.

This book brings together the information gathered from surveys of a variety of novel materials used for counter electrodes while improving the efficiency of DSSCs. Counter electrodes undertake three functions: (i) the oxidized dye is reduced by accepting electrons at the surface of counter electrodes, (ii) the dye transmits the electron into the cell after collecting them from the external circuit, and (iii) the mirror images of the counter electrodes reflect the unabsorbed photons from the cells to enhance the utilization of sunlight. Nowadays, such novel counter electrode materials have been developed that can not only perform the above three basic functions but also have excellent properties that are better than the standard platinum-based counter electrodes. Counter electrodes can be made of platinum, carbon materials, metal materials, conducting polymers and their composites, and transition metal compounds.

The book contains nine chapters on the fundamentals and materials used for counter electrodes. It aids to upgrade the knowledge of those working in multidisciplinary areas, such as physics, chemistry, nanoscience, energy science, and materials science. It can also be used for teaching and as a reference book by students of both bachelor's and master's degree programs. The first chapter deals with the functions of counter electrodes and their characterizations. Chapters 2 to 5 elaborate on metal oxides, carbides, nitrides, and sulphides as counter electrodes, and Chapter 6 discusses counter electrodes based on chalcogenides. The next two chapters elaborate on carbon-based counter electrodes such as carbon nanotubes and nanofibers and their composites. The final chapter gives information about quantum dots–based counter electrodes.

Last but not the least, we would like to express our thanks and gratitude to the chapter authors for generously sharing their knowledge on counter electrodes used in dye-sensitized solar cells for the benefit of our community.

Alagarsamy Pandikumar

Ramesh Mohan

Kandasamy Jothivnekatachalam

January 2021

Chapter 1

Functions of a Counter Electrode in Dye-Sensitized Solar Cells

C. R. Kalaiselvi,[a] T. S. Senthil,[a] and S. Kalpana[b]
[a]*Department of Physics, Erode Sengunthar Engineering College, Thudupathi, Erode 638476, Tamil Nadu, India*
[b]*Department of Physics, Academy of Maritime Education and Training, Chennai 603112, Tamil Nadu, India*
tssenthi@gmail.com

1.1 Introduction

Dye-sensitized solar cells (DSSCs) are photonic devices that convert visible light into electricity and are based on a porous thin film of a wide bandgap semiconductor oxide modified by dye molecules. In DSSCs, the enhancement of light absorption is due to their increased surface area and porosity of the film [1]. The four major elements that constitute the structure of DSSCs are the transparent conducting and counter conducting electrodes, the nanostructured wide bandgap semiconducting layer, the dye molecules (sensitizer), and the electrolyte [2]. In this cell, the anode is the electrode on which the metal oxide semiconductor is deposited and called photoanode as

Counter Electrode for Dye-Sensitized Solar Cells
Edited by Alagarsamy Pandikumar, Ramesh Mohan, and Kandasamy Jothivnekatachalam

ISBN 978-981-4877-38-1 (Hardcover), 978-1-003-11077-4 (eBook)
www.jennystanford.com

sunlight comes out from this anode. The cathode is the electrode on which platinum and other conducting materials are deposited and is more often called counter electrode (CE) because it exists relative to the anode and a more important reason is that Gratzel first used "counter electrode" in his pioneering paper in *Nature* in 1991 [3].

In DSSCs, the heart of the system is mesoporous TiO_2 (anatase) nanoparticles or nanostructures deposited on FTO (fluorine-doped tin oxide) substrate over which the photosensitizer (the dye) is coated. The schematic view of DSSCs is shown in Fig. 1.1. Photoexcitation of electrons occurs from the HOMO (highest occupied molecular orbital) to the LUMO (lowest unoccupied molecular orbital) of the dye, followed by the injection of electrons from the LUMO level to the conduction band (CB) of the semiconductor, leaving the dye in an oxidized state. The oxidized dye is restored to its original state by the transfer of electrons from the electrolyte, and it is called as the dye regeneration process. Regeneration of the dye by the iodide intercepts the recapture of the injected electron in the CB by the oxidized dye. The triiodide ions (I^{3-}) formed by the oxidation of iodide (I^-) diffuse through the counter electrode and the regenerative cycle is completed by the conversion of I^{3-} to I^- ions. Hence, the electrical circuit is completed by the migration of electrons to the external circuit. Under light illumination, a voltage is generated corresponding to the difference in Fermi level of the electron in the metal oxide and the redox potential of the electrolyte [4].

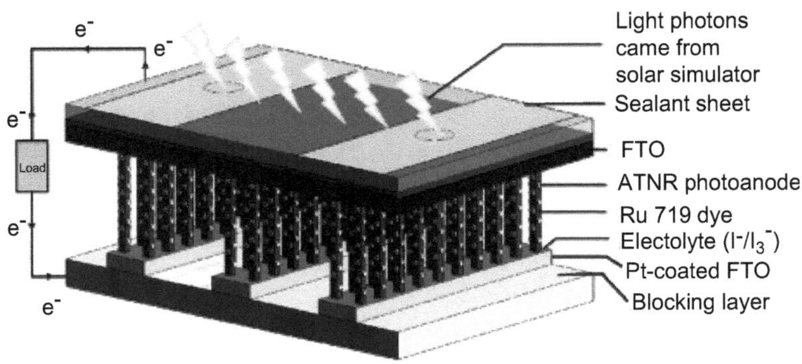

Figure 1.1 Schematic view of DSSCs. Reprinted from Ref. [35], Copyright 2018, with permission from Elsevier B.V.

Platinum (Pt) is used commonly as a counter electrode because Pt has high catalysis for electrolyte reduction. Even though DSSC with a Pt counter electrode achieved high energy conversion efficiency up to 11.1% [5], the cost issue of Pt reduces the fascination of DSSC development. Instead of Pt, carbon materials are attractive owing to their low cost, high conductivity, corrosion resistance to iodide ions, and high catalysis for I^{3-} reduction. Z. Huang et al. [6] discussed about the carbon materials such as hard carbon spherule, porous carbon, mesoporous carbon, and carbon powder used for DSSCs as the counter electrodes in the previous studies. Moreover, the conversion efficiencies of DSSCs with carbon materials were less than those with the Pt counter electrode. Therefore, further development in the manufacturing of the carbon counter electrodes is the greatest challenge for researchers.

Carbon materials are considered to be an excellent substitute for Pt owing to their good reduction ability to triiodine ions in electrolyte. J. Wu et al. [7] reviewed various kinds of carbon, such as activated carbon, carbon nanotubes (CNTs), graphite, hard carbon spherules, and carbon black, and their advantages of high electrical conductivity, chemical stability, and high surface area which are attractive features for the replacement of Pt as CE in DSSCs. Among the various counter electrodes such as metal alloy, carbon, polymer, transition metal compound, and hybrid, this chapter panoramically reviews the carbon-based CEs particularly on CNTs, which are of great significance in enhancing the efficiency of DSSCs. This chapter highlights the functions of counter electrode and its characterizations.

1.2 Counter Electrode and Its Role in DSSCs

The counter electrode (also known as auxiliary electrode) is an electrode that is used to close the current circuit in the electrochemical cell. It is usually made of an inert material (e.g., Pt, Au, graphite, glassy carbon) and usually it does not participate in the electrochemical reaction. In the paper by S. Yun et al., it is discussed that the counter electrode (CE) in DSSCs performs three functions [5, 8]: (i) as a catalyst, it promotes the completion of process, that is, the oxidized redox couple is reduced by accepting electrons at the surface of the CE, and the oxidized dye is reduced by collecting electrons via ionic transport materials in solid state DSSCs. (ii) As a positive electrode of primary cells, it collects electrons from the external circuit and transmits them into the cell. Thus, the ultimate function of the CE is to return electrons from the external load back into the "circulation" within the cell. (iii) As a mirror, it reflects the unabsorbed light from the cell back to the cell to enhance the better utilization of sunlight.

Apart from the basic functions stated above, J. Wu et al. [9] pointed out that an optimal CE should possess the qualities of having high catalytic activity, high conductivity, high reflectivity, low cost, high surface area, porous nature, optimum thickness, chemical, electrochemical and mechanical stability, chemical corrosion resistance, energy level that will match the potential of the redox couple electrolyte, good adhesivity with transparent conductive oxide (TCO), and so on. Further, an effective CE should have good conductivity and exhibit high electrocatalytic activity for reduction of the redox couple. The reduction reaction ($I^{3-} + 2e^- \leftrightarrow 3I^-$) on naked ITO (indium tin oxide) or FTO glass is extremely slow. Thus, in order to minimize the charge transfer overpotential, the catalytic materials should be coated on conducting substrates in order to speed up the reaction of the system.

O. Topon et al. [10] stated that the counter electrode of DSSCs serves an essential role in the reduction of the electrolyte oxidized at the photovoltaic electrode. They also discussed that the counter electrode most often utilized is platinum from the perspective of reduction capability and corrosion resistance to the electrolyte. However, the platinum has still issues of insufficient corrosion resistance to the iodide electrolyte and its cost. Hence, the carbon materials are potential alternative to it.

The major role played by CE is that it acts as a catalyst by reducing the redox species, which are the mediators for regenerating the sensitizer (dye) after the electron injection or for collecting the hole from the hole-transporting materials in a solid state DSSCs. Much of the research work in DSSC is focused on boosting the short-circuit current density (J_{SC}), open-circuit voltage (V_{OC}), and fill factor (FF) for increasing the efficiency. It is pointed out by S. Hao et al. [11] that Pt coated on FTO is used commonly as a CE in DSSCs. By improving the CE material, FF of the cell is raised, which is influenced by the series resistance (R_S) of the cell which is related to the slope of the tangent line to the current density and voltage (JV) curve at V_{oc}. J. D. Roy-Mayhew et al. [12] discussed that the reactions at the CE mainly depend on the redox species used to transfer electron between the photoanode and the CE. Usually I^-/I^{3-} couple is used as the redox mediator in DSSCs. The I^{3-} produced due to regeneration of the dye is reduced at the counter electrode by the following reaction:

$$I^{3-} + 2e^- \leftrightarrow 3I^-$$

A CE material in DSSC should possess high catalytic activity and stability towards the electrolyte used in the cell. Pt is a standard material of choice in DSSCs as a CE due to its attractive properties such as high catalytic activity, excellent stability towards the iodide redox species, and so on.

1.3 Requirements of Counter Electrode

The CE has important effects on the photovoltaic parameter of DSSCs. The theoretical maximum photovoltage of DSSCs is determined by the energy difference between the redox potential of the mediator and the Fermi level of the metal oxide semiconductor on the photoanode and it can be obtained only at the zero current. Under load, however, the output voltage is usually less than the open-circuit voltage and this voltage loss comes from the overall overpotential of the CE, which comes from the delivery of current through the electrolyte and through the electrolyte/counter electrode interface [13].

An effective CE should have good conductivity and exhibit high electrocatalytic activity for reduction of the redox couple as stated by J. Wu et al. [9]. However, in order to minimize the charge transfer overpotential, catalytic materials should be coated on conducting substrates to speed up the reaction. Novel CE materials present excellent properties, even better than those of platinum, which are significant to promote the industrialization of DSSCs. According to the material composition, CEs can be made of platinum, other metal materials, carbon materials, transition metal compounds, conductive polymers, and composites. As per the literature reviews, carbon materials are the most commonly used CE materials in DSSCs, and published paper share of the total articles on CEs constitutes 23% and the patents account for 47% of the total.

1.4 Limitations of Counter Electrode

Until now, great progresses have been made in the CE catalysts in DSSCs. As far as the CE catalysts are concerned, they should possess (i) better catalytic activity, (ii) better electrical conductivity, (iii) better chemical stability, (iv) large specific surface area, (v) high corrosion resistance, (vi) better matching for high-performance DSSCs, which must be certain while considered in the design of Pt-free catalysts to maximize the function of CEs. Every aspect of consideration has pros and cons as CE most commonly used is Pt because it possesses a high catalytic activity, but its cost limits its usage. Conventional carbon is perhaps an alternative to it; however, a large amount of carbon is required to reach the full performance level of Pt electrodes, which results in a bulky and opaque DSSC device. Conventional carbon CE catalysts also suffer from poor adhesion between the CE catalytic layer and the FTO substrate. By contrast, CNTs and graphene are suitable catalytic materials for I^{3-} reduction due to their distinctive structures and properties and they could be an alternative for Pt-free CE catalysts in DSSCs. However, the efficiency achieved will be quite low compared to Pt CE [14].

1.5 Characterization of Counter Electrode

1.5.1 Electron Impedance Spectra and Nyquist Plot

Electrochemical impedance spectroscopy (EIS) is a powerful tool for studying the kinetics of charge transport and electron–hole recombination and for evaluating the electrocatalytic activity of counter electrodes in DSSCs [9]. Q. Wang et al. [15] performed the study regarding the impedance spectra to investigate electronic and ionic processes in DSSCs. They have done different processes for the cell under dark or bright condition. Initially at open-circuit voltage and in sunlight, there is no net current flowing through the cell. All the injected electrons are recaptured by I^{3-} before being extracted to the external circuit. Meanwhile, the oxidized dye is regenerated by I^-. As a result of this, the absorbed photon energy is converted to heat through the two coupled redox cycles involving sensitized electron injection, dye regeneration, and electron recapture by I^{3-}. The counter electrode is kept at equilibrium, since there is no net current flowing through it. However, in the dark condition under forward bias, electrons are transported through the mesoscopic TiO_2 network and react with I^{3-}. At the same time, I^- is oxidized to I^{3-} at the counter electrode. The net current density can be large depending on the applied bias voltage.

As stated by W. J. Lee et al. [16], the faster degradation of carbon counter electrode DSSC in comparison with Pt counter electrode device prepared at the similar condition may be related to moisture content in carbon. Therefore, prior to applying such high performance carbon counter electrodes in DSSCs, water molecule must be removed from the pores in the carbon. Continuous variation of device performance raises the concern about water and oxygen permeability. During their study, both V_{oc} and FF gradually increased while J_{sc} decreased. Although a similar trend was observed in the Pt counter electrode DSSC, the magnitude of J_{sc} degradation in carbon counter electrode device was more evident, and hence device retains 84% of initial day performance. Dark current–voltage and electrochemical impedance spectroscopy measurements confirm the robustness of the carbon counter electrode in the organic liquid electrolyte.

EIS data are often expressed in a Nyquist plot or a Bode plot [9]. In a complete Nyquist plot for DSSCs, there are usually three semicircles. The starting point is R_s, mainly caused by the conductive substrate, connecting wires, and so on; the first semicircle (high frequency) refers to the charge transfer resistance (R_{ct}) at the CE/electrolyte interface, the second semicircle (middle frequency) corresponds to the R_{ct} at the anode/dye/electrolyte interface, and the third semicircle (low frequency) is attributed to the redox species diffusion resistance (Z_W). Owing to the small distance between two electrodes and the low viscosity of the electrolyte, the third semicircle is often not observed. Impedance spectra of DSSC with N719 sensitizer model is shown in Fig. 1.2.

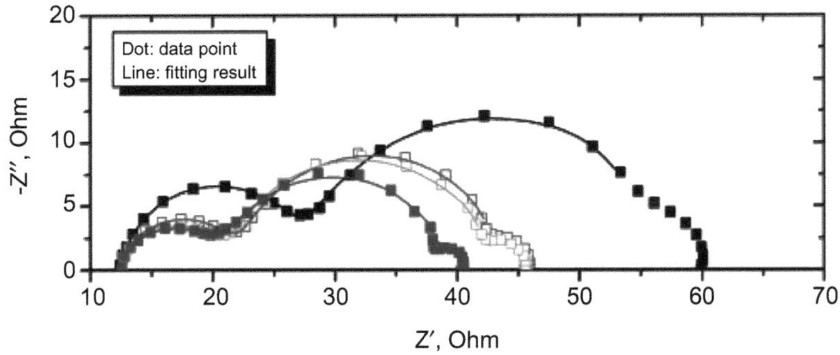

Figure 1.2 Impedance spectra of DSSC using N719 sensitizer. Reprinted with permission from Ref. [15], Copyright 2005, American Chemical Society.

1.5.2 Cyclic Voltammetry

Electrochemical methods are used to characterize all the components of DSSCs. They give significant information regarding the energy levels of components, the reversibility of electrochemical reactions, and the kinetics of electrochemical processes. Cyclic voltammetry (CV) is an important tool for investigating counter electrodes. A three-electrode system is widely used to carry out CV. It consists of a working electrode, a reference electrode, and a counter (auxiliary) electrode [9]. The potential is swept at a constant rate and reversed at a certain point, while the current is monitored continuously. The potential is measured between the working electrode and the reference electrode, while the current is measured between the working electrode and the counter electrode. These data are plotted as current (*I*) versus applied potential (*E*, often referred to as "potential"). A typical cyclic curve as shown in Fig. 1.3 signifies that the waveform of even reversible redox couples is complex owing to the combined effects of polarization, diffusion, and the rate of electron transfer. In DSSCs, the iodide/triiodide couple is often used as a redox mediator in liquid electrolytes. The CV curves in iodide/triiodide electrolytes often contain two pairs of redox peaks.

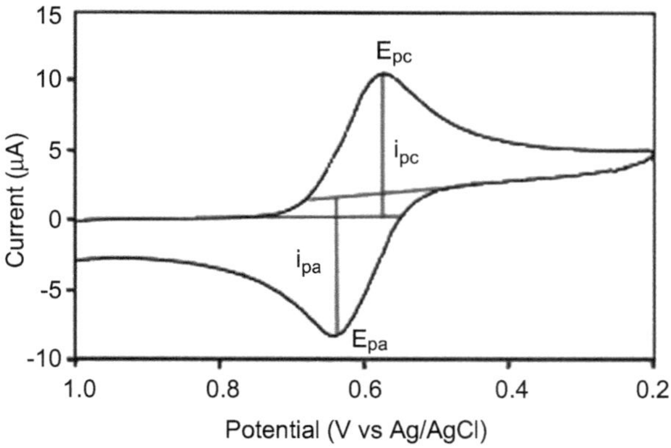

Figure 1.3 Typical cyclic voltammogram where i_{pc} and i_{pa} show the peak cathodic and the anodic current, respectively, for a reversible reaction. Reproduced from Ref. [7], Copyright 2017, by permission of The Royal Society of Chemistry.

1.6 Materials Used as Counter Electrode

1.6.1 Pt Electrode

S. Hwang et. al. [14] reported that Pt has long been known as the most effective catalyst material for the counter electrode in DSSCs, owing to its excellent electrocatalytic activity for the reduction of triiodide. They also pointed out the various techniques employed in the preparation of Pt counter electrodes, such as thermal decomposition, electrodeposition, spin coating, physical vapor deposition, electrochemical deposition, and sputtering. They also reviewed that the highest catalytic activity and the lowest R_{ct} of the Pt counter electrode were achieved at a heating rate of 1.2 °C/min. Further, this heating rate provided a uniform distribution of Pt nanoparticles on a FTO glass substrate. The best Pt counter electrode in a DSSC exhibited an energy conversion efficiency of 9.3%. Although the most common methods used are thermal decomposition and sputtering for the preparation of Pt films, they require inherently complicated processes and large amounts of expensive Pt source. Although the performance of Pt electrodes employed by DSSCs can be improved by controlling the structures and morphologies, the reality is that such Pt nanostructure-based solar cells may not be commercialized with large-scale designs. Regarding these issues, various cost-effective materials have been explored to substitute Pt films.

1.6.2 Graphene Electrode

Cost efficiency to replace the Pt counter electrode has led to the investigation of carbonaceous materials, including carbon black, graphene, CNTs, and their composites for DSSCs. Graphene is a quite attractive material to replace Pt due to its excellent conductivity, corrosion resistance, low cost, and ability to decrease the value of R_{ct}. Kaniyoor and Ramaprabhu [17] showed that the value of R_{ct} for thermally exfoliated graphene (TEG) films was 11.7 Ω cm^2, which is very close to that of Pt electrodes (6.5 Ω cm^2). Also, DSSCs fabricated with TEG as a counter electrode exhibited a power conversion efficiency of about 2.8%, comparable to that (about 3.4%) of Pt-based DSSCs. Furthermore, similar considerations were made by Zhang et al. [18] that the R_{ct} value of graphene nanosheet (GN) films was also close to that of Pt and the cell showed a J_{sc} of 17 mA/cm^2, an V_{oc} of 0.747 V, and a FF of 53.62% under the illumination of 100 mW/cm^2, leading to a highest efficiency of 6.8%. The difference in the achieved efficiency is mainly attributed to the different techniques involved in the preparation of the graphene film and the DSSC fabrication. Further work done by S. H. Huh et al. [19] reported that reduced graphene oxide (RGO) functionalized with –NHCO– groups exhibited a higher catalytic activity than the original RGO. However, it is uncertain which specific oxygen-containing functional groups in the RGO sheets are mainly responsible for the electrocatalysis of I^{3-}/I^-. In addition, DSSC exhibited a low conversion efficiency equal to 40% of that of a Pt-based cell that was made from partially RGO. X. Gao et al. [20] signified that the reason beyond the lowest efficiency was attributed to too many oxygen-containing groups, which decreased the electrical conductivity of graphene. Therefore, care should be taken to have equivalence between the electrical conductivity and the electrocatalytic activity of graphene to optimize the DSSC by functionalization.

1.6.3 Carbon Nanotubes as a Counter Electrode

Carbon nanotubes have been a fascinating material for counter electrodes since their characteristics such as conductivity and specific surface area are high when compared with other carbon materials. It is also a highly conductive material, which decreases the R_{ct} value of films. The films with low R_{ct} values exhibit good electron-transfer kinetics and can improve the performance of DSSCs. E. Ramasamy et al. [21] investigated the use of low conduction substrates such as an FTO glass and a polytetrafluoroethylene (PTFE) membrane with CNT counter electrodes in the performance of DSSC and suggested that CNT during I^{3-} reduction showed low impedance. They also suggested that the high surface area and low sheet resistance of CNT make CNTs a promising replacement for highly energy-efficient DSSCs.

CNTs are graphene sheets that have been rolled in the cylindrical form and capped with half of the fullerene structure. Further, it possesses nanoscale dimensions with very high aspect ratios that provide unusual multifunctional properties. Generally, CNTs can be classified into single-walled CNTs (SWCNTs), double-walled CNTs (DWCNTs), and multi-walled CNTs (MWCNTs), depending on the number of walls. Suzuki et al. [22] narrated that with the very high surface area and low sheet resistance, a SWCNT film was first used as the counter electrode in a DSSCs and its performance was compared with the Pt counter electrode. The PV (photovoltaic) efficiency of only 3.5% was achieved with SWCNT-based DSSCs, which was quite low, whereas for the conventional type cell, it was 5.4%. Later, in one of the inspiring works done by Lee et al. [23], bamboo-structured MWCNTs were used as an electrocatalyst in the counter electrode of a DSSC. In that, the R_{ct} value of the MWCNT film was measured to be 0.82 Wcm^2, which was two times lower than that of the screen-printed Pt electrode (1.8 Wcm^2). The defect-rich edge lanes present in the bamboo-structured CNTs are responsible for the lowest value of R_{ct}. Due to this fact, DSSCs fabricated with MWCNT counter electrodes achieved an energy conversion efficiency of 7.67%, whereas for the conventional Pt-based cell, it was 7.83%.

1.6.3.1 Preparation techniques

To date, various techniques have been developed and adopted to deposit CNTs on the substrate, such as tape cast, electrophoretic deposition, screen printing, chemical vapor deposition (CVD), spray coating, and spin coating. Among these strategies, electrophoretic deposition, doctor blading, vacuum filtration, spin coating, and screen printing are considered to be simple and cost-effective, since they employ solution processable samples [24]. J. D. Roy-Mayhew et al. [4] discussed that dispersing pristine CNTs in base liquids (water, ethanol, ethylene glycol, etc.) was very difficult on the account of their agglomeration and entanglement caused by the long-winding shape and hydrophobic surfaces of CNTs. In order to overcome this barrier, an acid oxidation process with nitric acid (HNO_3) and sulfuric acid (H_2SO_4) was used. Further, during this oxidation process, functional groups such as hydroxyl (–OH), carbonyl (C=O), and carboxyl (–COOH) were created on the CNT surface. Thus, chemically treated or functionalized CNTs can easily be dispersed in base fluids. Therefore, these oxygen-containing functional groups on CNTs enhanced the electrocatalytic activity for the reduction of the redox electrolyte.

Trancik et al. [25] recorded that the electrocatalytic activity of CNTs was improved by the introduction of defects on nanotubes. They also found that the oxidized (ozone-treated) CNTs can serve as catalytic electrodes for DSSCs. Although highly functionalized CNTs improved the electrocatalytic activity, they still proved to have poor conductivity compared to pristine CNTs. Until now, the functionalization level of CNTs that afford the best conditions for the performance of solar cells has not been fully understood. Therefore, the only way of finding an optimized functionalization level of CNTs was by adjusting the experimental conditions such as reacting time, samples ratio, and concentration of acids, which showed a great importance in the overall performance of DSSCs. Zhang et al. [26] fabricated three different DSSCs based on SWCNTs, DWCNTs, and MWCNTs and compared their PV performances with the same conditions and the sequence of PV efficiencies was 8.03 > 7.61 > 7.06% for DWCNTs > SWCNTs > MWCNTs, respectively. The PV efficiencies of DSSCs fabricated with three different counter electrodes such as SWCNTs, MWCNTs, and Pt films are shown in Fig. 1.4 [36].

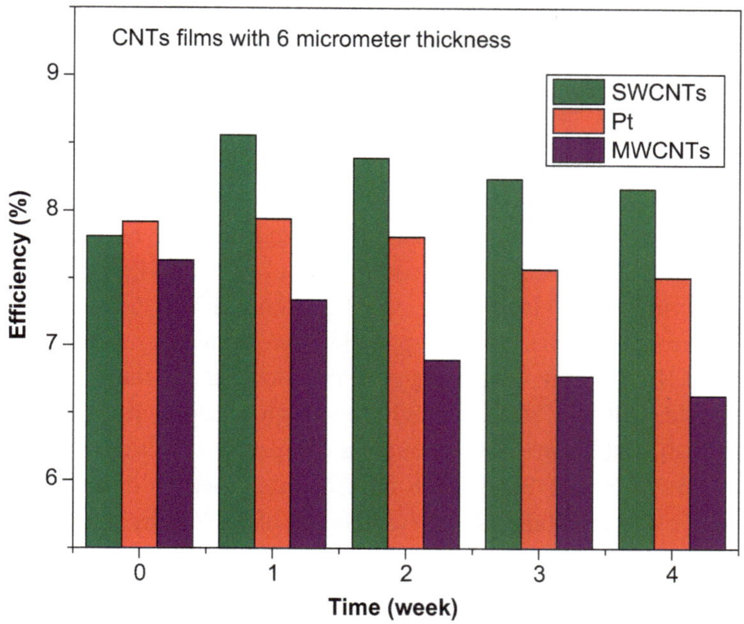

Figure 1.4 PV efficiency stability of DSSCs fabricated with three different counter electrodes, namely, gel-coated binder-free SWCNTs, MWCNTs, and Pt films. Data points are adapted from Ref. [36].

The well-aligned CNT counter electrode possesses many features that can be beneficial to achieve high PV performance in DSSCs. J. G. Nam et al. [27] showed the advantages of CNT counter electrode as: (1) having better adhesion between CNTs and substrate, (2) extremely high surface-area-to-volume ratio with a great geometrical order, (3) fast electron transfer through the nanotubes (one dimensional), and (4) excellent electrocatalytic activity. Further, they concluded that although CNT-based films possess lower R_{ct} values and faster reaction rates compared to Pt films, the energy conversion efficiency of DSSCs with CNT counter electrodes (7.03%) was slightly lower than that of Pt electrode-based DSSCs (7.88%). However, such a slight difference (between CNT and Pt electrodes) in the PV performance of DSSCs may be ignorable, because the CNT-based DSSCs have both a lower cost and superior chemical stability compared to the Pt-based devices.

1.6.4 CNT-Based Composites

1.6.4.1 CNT–polymer composites

CNT/polymer composite materials have been intensively exploited in energy-related applications due to their multifunctional properties. Various types of polymeric structures such as poly(styrenesulfonate) (PSS), poly(3,4-ethylenedioxythiophene) (PEDOT), PEDOT:PSS, poly(vinyl chloride) (PVC), polythiophene (PTh), polypropylene, polyaniline (PANI), and polypyrrole have been utilized to support the CNT counter electrode of DSSCs, owing to their low cost, excellent electrocatalytic activity, and high conductivity [14]. Fan et al. [28] first pointed out about the composite structure composed of MWCNTs and PEDOT:PSS as a counter electrode material for DSSCs and it exhibited a PV efficiency of 6.5%. The authors also described that the high performance of the DSSC obtained with this composite structure is attributed to rapid electron transfer at the counter electrode caused by π–π conjugation between MWCNTs and PEDOT:PSS.

1.6.4.2 CNT–metal composites

In order to reduce the Pt content in the counter electrode, one of the strategies followed is the use of composite materials composed of conductive CNTs and a lesser amount of catalytic Pt particles. Chen et al. [29] synthesized CNT/Pt nanohybrids using an ionic liquid-assisted sonochemical method and deposited them on the substrate through a simple electrospray method. With the concentration, deposition time, and flow rate of the solution, they adjusted the film thickness. From their result, it was revealed that DSSCs fabricated based on a MWCNT/Pt film with 100 nm thickness exhibited a PV efficiency of 8.81%, which was close to that (8.84%) of the Pt-only counter electrode in DSSCs, despite the amount of Pt precursor used in the CNs/Pt hybrid being significantly lower than that used in the reference cell. This high efficiency achieved by the MWCNT/Pt-film-based DSSC was attributed to the excellent electrocatalytic property of Pt and the high conductivity of CNTs. Tai et al. [30] showed that the transition metal compounds also had a remarkable catalytic activity (similar to Pt), but their low conductivity is the main drawback for them to be used as ideal counter electrode materials for DSSCs. Therefore, an efficient way to overcome this issue is to incorporate such inorganic compounds with other conductive materials such as titanium nitride (TiN), molybdenum nitride (MoN), molybdenum disulfide (MoS_2), dinickel phosphide (Ni_2P), tungsten disulfide (WS_2), nickel sulfide (NiS), and cobalt sulfide (CoS) for use in DSSC cathodes. For instance, Li et al. [31] prepared TiN–CNT composite-based counter electrodes of DSSCs to act as a high electron transporter and a good catalyst. Among all metal composites, the NiS–MWCNT/Ti film showed the highest catalytic activity and conductivity. This high performance of the composite film was attributed to an excellent synergistic effect between the NiS and MWCNTs. It yielded an energy conversion efficiency of 7.9%, which was significantly higher than that (6.36%) of the cell assembled with Pt/Ti film.

1.6.4.3 CNT–graphene composites

Yang et al. noted that the aggregation of CNTs significantly hampers their charge transport property, thus resulting in a lower efficiency than expected [32]. One possible approach to decrease the aggregation of CNTs is to incorporate them with the 2D carbon material graphene. Moreover, graphene materials have unique properties, such as superior electrical conductivity, excellent catalytic property, and very high surface area. Furthermore, defect-rich graphene structures have been demonstrated to be promising alternatives to expensive Pt for use in the counter electrode of DSSCs. Choi et al. [33] prepared a MWCNT–graphene composite counter electrode for DSSCs. In that the graphene layers were first deposited on a SiO_2/Si substrate by using a drop-casting method and by using a CVD technique, MWCNTs were vertically aligned on graphene layers. The device fabricated showed a power conversion efficiency of only 3.0%. Then, this low PV performance of the DSSC with the MWCNT–graphene film was further improved to 6.17% by other researchers. Zhu et al. [34] noted six different MWCNT–graphene films by varying the content of CNTs (0, 20, 40, 60, 80, and 100 wt%) in the composite. It was found by them that the CNT content in the nanocomposite had a vital role in the performance of the cell. The initial improvement in the cell performance was assigned to the excellent network structure between graphene sheets and CNTs. However, the decreased efficiency with further increase in the CNT content was associated with the surface area of CNTs. Indeed, the highest efficiency of 6.17% was achieved by the DSSC with a 60 wt% CNTs/40 wt% graphene counter electrode. Table 1.1 shows the PV characteristics of different counter electrodes used in DSSCs.

Table 1.1 PV characteristics of Pt-, graphene-, and CNT-based counter electrodes

Counter electrode	Short-circuit current density (J_{sc}) (mA/cm^2)	Open-circuit voltage (V_{oc}) (V)	Fill factor (FF)	Efficiency (η) %	References
Pt	18.67	0.713	0.747	9.95	10
Graphene-400°C	16.988	0.747	53.62	6.81	18
SWCNT	17.39	0.711	0.727	8.99	10
MWCNT	8.41	0.720	0.71	7.96	16
MWCNTs/PSSNa	16.9	0.710	0.59	7.03	14
NiS-MWCNTs/Ti	14.18	0.75	0.74	7.90	14

1.7 Conclusion

Though a vast research done on CNTs as a counter electrode in DSSCs already exists as detailed in this chapter, there are still many avenues of investigation regarding the efficiency to be pursued. If these avenues are explored, surely CNTs will play a key role in the development of DSSCs, which will provide energy at competitive costs compared to current energy production approaches. Newly developed processes are continually being reported in future and they will provide an opportunity for innovation in research and ensure that the production does not enhance the greenhouse effect on environment.

References

1. S. Suhaimi, M. M. Shahimin, Z. A. Alahmed, J. Chysky and A. H. Reshak, Materials for enhanced dye-sensitized solar cell performance: electrochemical application, *Int. J. Electrochem. Sci.*, 10: 2859–2871, 2015.
2. K. E. Jasim, Dye sensitized solar cells – working principles, challenges and opportunities, *Solar Cells – Dye-Sensitized Devices*, ISBN 978-953-307-735-2, 2011.
3. D. A. Cusano, CdTe solar cells and photovoltaic heterojunctions in II–VI compounds, *Solid-State Electron.*, 6: 217–232, 1963.
4. J. D. Roy-Mayhew, D. J. Bozym, C. Punckt and I. A. Aksay, Functionalized graphene as a catalytic counter electrode in dye-sensitized solar cells, *ACS Nano*, 4: 6203–6211, 2010.
5. S. Thomas, T. Deepak, G. Anjusree, T. Arun, S. Naira and A. Nair, A review on counter electrode materials in dye-sensitized solar cells, *J. Mater. Chem. A*, 2: 4474–4490, 2014.
6. Z. Huang, X. Liu, K. Li, D. Li, Y. Luo, H. Li, W. Song, L. Chen and Q. Meng, Application of carbon materials as counter electrodes of dye-sensitized solar cells, *Electrochem. Commun.*, 9: 596–598, 2007.
7. J. Wu, Z. Lan, J. Lin, M. Huang, Y. Huang, L. Fan, G. Luo, Y. Lin, Y. Xie and Y. Wei, Counter electrodes in dye-sensitized solar cells, *Chem. Soc. Rev.*, 46: 5975–6023, 2017.
8. S. Yun, A. Hagfeldt and T. Ma, Pt-free counter electrode for dye-sensitized solar cells with high efficiency, *Adv. Mater.*, 26: 6210–6237, 2014.
9. J. Wu, Y. Li, Q. Tang, G. Yue, J. Lin, M. Huang and L. Meng, Bifacial dye-sensitized solar cells: a strategy to enhance overall efficiency based on transparent polyaniline electrode, *Sci. Rep.*, 4: 4028, 2014.
10. O. Topon, D. Matsumoto and M. Inaguma, Carbon nanotubes counter electrode for dye-sensitized solar cell, *Fujikura Tech. Rev.*, 47–50, 2011.
11. S. Hao, J. Wu, J. Lin and Y. Huang, Modification of photocathode of dye-sensitized nanocrystalline solar cell with platinum by vacuum coating,

thermal decomposition and electroplating, *Compos. Interfaces*, 13: 899–909, 2006.

12. J. D. Roy-Mayhew, D. J. Bozym, C. Punckt and I. A. Aksay, Functionalized graphene as a catalytic counter electrode in dye-sensitized solar cells, *ACS Nano*, 4: 6203, 2010.
13. N. Papageorgiou, Counter-electrode function in nanocrystalline photoelectrochemical cell configurations, *Coord. Chem. Rev.*, 248: 1421–1446, 2004.
14. S. Hwang, M. Batmunkh, M. J. Nine, H. Chung and H. Jeong, Dye-sensitized solar cell counter electrodes based on carbon nanotubes, *ChemPhysChem*, 15: 1–14, 2014.
15. Q. Wang, J.-E. Moser and M. Gratzel, Electrochemical impedance spectroscopic analysis of dye-sensitized solar cells, *J. Phys. Chem. B*, 109: 14945–14953, 2005.
16. W. J. Lee, E. Ramasamy, D. Y. Lee and J. S. Song, Performance variation of carbon counter electrode based dye-sensitized solar cell, *Sol. Energy Mater. Sol. Cells*, 92: 814–818, 2008.
17. A. Kaniyoor and S. Ramaprabhu, Thermally exfoliated graphene based counter electrode for low cost dye sensitized solar cells, *J. Appl. Phys.*, 109: 124308, 2011.
18. D. W. Zhang, X. D. Li, H. B. Li, S. Chen, Z. Sun, X. J. Yin and S. M. Huang, Graphene-based counter electrode for dye-sensitized solar cells, *Carbon*, 49: 5382–5388, 2011.
19. W. Lu, J.-B. Baek and L. Dai (Eds.), *Carbon Nanomaterials for Advanced Energy Systems: Advances in Materials Synthesis and Device Applications*, 2015.
20. X. Gao, J. Jang and S. Nagase, Hydrazine and thermal reduction of graphene oxide: reaction mechanisms, product structures, and reaction design, *J. Phys. Chem. C*, 114: 832–842, 2010.
21. E. Ramasamy, W. J. Lee, D. Y. Lee and J. S. Song, Spray coated multi-wall carbon nanotube counter electrode for tri-iodide (I^{3-}) reduction in dye-sensitized solar cells, *Electrochem. Commun.*, 10: 1087–1089, 2008.
22. K. Suzuki, M. Yamaguchi, M. Kumaga and S. Yanagida, Application of carbon nanotubes to counter electrodes of dye-sensitized solar cells, *Chem. Lett.*, 32: 28, 2003.
23. W. J. Lee, E. Ramasamy, D. Y. Lee and J. S. Song, Efficient dye-sensitized solar cells with catalytic multiwall carbon nanotube counter electrodes, *ACS Appl. Mater. Interfaces*, 1: 1145, 2009.

24. J. Y. Roh, Y. H. Kim and C. S. Lee, Synthesis of MWNTs using thermal chemical vapor deposition for the application of a counter electrode for DSSCs, *Curr. Appl. Phys.*, 11: S69, 2011.

25. J. E. Trancik, S. C. Barton, and J. Hone, Transparent and catalytic carbon nanotube films, *Nano Lett.*, 8: 982, 2008.

26. D. Zhang, X. Li, S. Chen, Z. Sun, X. J. Yin and S. Huang, Performance of dye-sensitized solar cells with various carbon nanotube counter electrodes, *Microchim. Acta*, 174: 73, 2011.

27. J. G. Nam, Y. J. Park, B. S. Kim and J. S. Lee, Enhancement of the efficiency of dye-sensitized solar cell by utilizing carbon nanotube counter electrode, *Scr. Mater.*, 62: 148, 2010.

28. B. Fan, X. Mei, K. Sun and J. Ouyang, Conducting polymer/carbon nanotube composite as counter electrode of dye-sensitized solar cells, *Appl. Phys. Lett.*, 93: 143103, 2008.

29. H. Y. Chen, J. Y. Liao, B. X. Lei, D. B. Kuang, Y. Fang and C. Y. Su, Highly catalytic carbon nanotube/Pt nanohybrid-based transparent counter electrode for efficient dye-sensitized solar cells, *Chem. Asian J.*, 7: 1795, 2012.

30. S. Y. Tai, C. J. Liu, S. W. Chou, F. S. S. Chien, J. Y. Lin and T. W. Lin, Facile synthesis of MoS_2/graphene nanocomposite with high catalytic activity toward triiodide reduction in dye-sensitized solar cells, *J. Mater. Chem.*, 22: 24753, 2012.

31. J. Song, G. R. Li, F. Y. Xiong, and X. P. Gao, Synergistic effect of molybdenum nitride and carbon nanotubes on electrocatalysis for dye-sensitized solar cells, *J. Mat. Chem.*, 122: 3735, 2010.

32. Z. Yang, T. Chen, R. He, G. Guan, H. Li, L. Qiu and H. Peng, Aligned carbon nanotube sheets for the electrodes of organic solar cells, *Adv. Mater.*, 23: 5436, 2011.

33. H. Choi, H. Kim, S. Hwang, W. Choi and M. Jeon, Dye-sensitized solar cells using graphene-based carbon nano composite as counter electrode, *Sol. Energy Mater. Sol. Cells*, 95: 323, 2011.

34. G. Zhu, L. Pan, T. Lu, T. Xu and Z. Sun, Electrophoretic deposition of reduced graphene-carbon nanotubes composite films as counter electrodes of dye-sensitized solar cells, *J. Mater. Chem.*, 21: 14869, 2011.

35. N. Sriharan, N. M. Ganesan, M. Kang, L. Kungumadevi and T. S. Senthil, Improved photoelectrical performance of single crystalline rutile TiO_2 nanorod arrays incorporating α-alumina for high efficiency dye-sensitized solar cell, *Mater. Lett.*, 237: 204–208, 2019.

36. X. Mei, S. J. Cho, B. Fan and J. Ouyang, High-performance dye-sensitized solar cells with gel-coated binder-free carbon nanotube films as counter electrode, *Nanotechnology*, 21: 395202, 2010.

Chapter 2

Trends in Metal Oxides Based Counter Electrode in Dye-Sensitized Solar Cells

Rajkumar C.[a] **and A. Arulraj**[b,c]
[a]*Department of Electronics and Communication, University of Allahabad, Prayagraj (Allahabad), India*
[b]*Graphene and Advanced 2D Materials Research Group (GAMRG), School of Science & Technology, Sunway University, Selangor, Malaysia*
[c]*University College of Engineering, Bharathidasan Institute of Technology (BIT) Campus, Anna University, Tiruchirappalli, India*
rajkumar.srec@gmail.com, rajkumarc@allduniv.ac.in

Solar cells are the prominent alternative energy resources in present era to replace the fossil-fuel-based energy resources. Solar cells using photovoltaic (PV) effect convert solar energy into electrical energy. The architecture of DSSCs (dye-sensitized solar cells) consists of several components, such as photoanode, electrolyte, and counter electrode (CE). The photoexcited electron from the dye reaches the external circuit through the photoanode and then the counter electrode collects the electron from an external circuit to catalyze the redox reduction in the electrolyte by reducing the overpotential. Therefore, counter electrodes play a significant role in boosting the performance of devices. So far, in counter electrodes, platinum (Pt) noble rare earth metal was used as the standard material because of its good catalytic property and long-term stability. However, it still

Counter Electrode for Dye-Sensitized Solar Cells
Edited by Alagarsamy Pandikumar, Ramesh Mohan, and Kandasamy Jothivnekatachalam

ISBN 978-981-4877-38-1 (Hardcover), 978-1-003-11077-4 (eBook)
www.jennystanford.com

lacks in transferring the technology because platinum is expensive. Hence, the researchers point out a different kind of materials possessing less expensive, easily available, long-term stability, and good catalyzing properties to replace the platinum metal. In this concern, metal oxides are found to be one of the best alternative materials to replace the conventional Pt counter electrode because of their cost-effectiveness and availability. Metal oxides have been easily synthesized using different approaches, thereby their structure, bandgap, and morphology can be tuned. For example, one-dimensional (1D) WO_2 and CuO nanorods hold an excellent catalytic activity close to the conventional Pt. Similarly, ZnO nanowire array composites exhibit higher efficiency than Pt. In addition to all the above, metal oxides help to facilitate electron transport to the electrolyte for improving the reduction process of triiodide ions by means of creating oxygen vacancy. So, in this chapter, preparation of different metal oxides, tuning the bandgap, morphological aspects, the role of defects, catalytic property, and the role of counter electrodes in DSSCs are briefly discussed.

2.1 Introduction

In universe, sustainable development is a major challenge for energy resources. Renewable energy is the most suitable sustainable energy resource. In renewable energy resources, the solar cell is the preferred one as compared to other renewable energy resources in the developed countries. Solar cells are the predominant alternative energy resources in the present era to replace fossil-fuel-based energy resources. A solar cell or PV cell is a device that converts sunlight into direct electricity. In another way, the solar cell can be denoted as a converter of solar light energy into electrical energy. DSSC is the third-generation solar cell as well as a major challenging solar cell for the conventional silicon solar cell due to its excellent properties, such as semitransparency, higher flexibility, a variety of colors, low-cost fabrication and environmental friendliness. It produces a robust increase in photoconversion efficiency (PCE) with a decrease in production cost for long-term development [1]. The architecture of DSSCs consists of several components, such as photoanode, electrolyte, and CE. The photoexcited electron from the dye reaches the external circuit through the photoanode and then the counter electrode collects the electron from an external circuit to catalyze the redox reduction in the electrolyte by reducing the overpotential. Therefore, counter electrodes play a significant role in boosting the performance of devices. Pt CE is mostly used in fabricating DSSCs because of its good catalytic activity and high stability. Pt is a highly expensive noble metal and has very less reserve material in the world. So there is a need to explore Pt-free counter electrodes with the same potential activity or higher than Pt [2, 3].

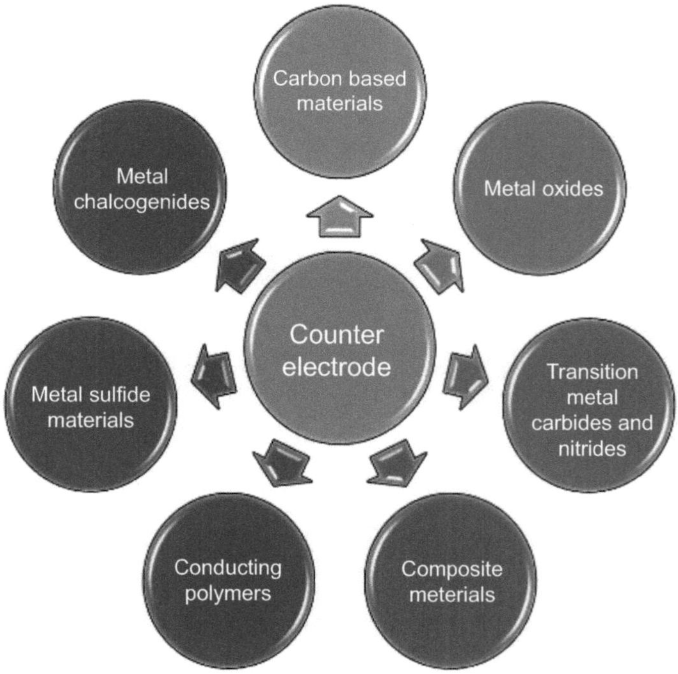

Figure 2.1 Different types of materials involved in the preparation of CE in DSSCs.

Hence, researchers point out different kinds of materials possessing less expensive, easily available, long-term stability, and good catalyzing properties to replace the platinum metal [4, 5]. Various studies are focusing on the development of different CE materials in DSSCs, such as carbon-based materials, metal oxides, transition metal carbides and nitrides, metal sulfides, metal chalcogenides, composites materials, and polymers (Fig. 2.1)

[6–11]. In these materials, metal oxides are found to be one of the best alternative materials to replace the conventional Pt counter electrode because of their cost-effectiveness and availability. Metal oxides have been easily synthesized using different approaches, thereby their structure, bandgap, and morphology can be tuned. In addition to all the above, metal oxides help to facilitate electron transport to the electrolyte for improving the reduction process of triiodide ions by means of creating oxygen vacancy.

2.2 Role of the Counter Electrode in DSSCs

There are three main functions of CE: (1) It acts as a catalyst to enhance the completion of process, that is, oxidized redox couple accepts electrons at the surface of CE and the oxidized dye is reduced by collecting the electrons through ionic transport materials in solid-state DSSCs. (2) The ultimate function of the CE is to collect the electrons from the external load and send them back into the circulation within the cell. (3) The CE act as mirror because the unabsorbed light is sent back to the cell to utilize the sunlight and increase or enhance the efficiency. In solid-state DSSCs, CE collects the hole from hole transport materials [12].

2.3 Basic Function and Optimal Qualities of CE

The following qualities are required to become optimal CE: high catalytic activity, high conductivity, high reflectivity, low cost, high surface area, porous nature, optimum thickness, electrochemical and mechanical stability, chemical corrosion resistance, energy level that matches the potential of the redox couple electrolyte, good adhesivity with TCO, and so on. Benchmark parameters for an ideal CE include 80% optical transparency at a wavelength of 550 nm, 2–3 Ω cm^2 charge transfer resistance (R_{ct}), and <20 Ω/sq sheet resistance (R_s). The overvoltage should be maintained as low as possible at the photocurrent density up to 20 mA/cm^2 [2]. However, it is not possible to synthesize a single material delivering all the required parameters. In carbon-based CE, the required R_s and R_{ct} can be achieved, but optical transparency is not ideal.

2.4 Metal Oxides and Their Necessity Towards CE

Metal oxides have been used for longer generation in technological and industrial sectors due to their diverse properties (electrical, optical, magnetic, etc.) combined with the overall characteristics of their hardness, thermal stability, and chemical resistance. The metallic compound formed with the composition of metal and oxygen in the form of oxide ion (O^{2-}) is called metal oxide (MO). MO formula is written with two words, where the first is the symbol of the metal with the oxidation number and the second is the oxygen symbol "O" with an oxidation number –2. Mostly MOs are basic in nature and many times they exist in solid form at room temperature. Basically, MOs are insoluble in water and produce salts with acids. For example, magnesium oxide reacts with hydrochloric acid to form magnesium chloride salt, that is, $MgO + 2HCl \rightarrow MgCl_2 + H_2O$. Some MOs are amphoteric or acidic in nature and produce salts with base in the neutralization reaction. For example, aluminium oxide reacts with sodium hydroxide to form sodium aluminate: $Al_2O_3 + 2NaOH \rightarrow 2NaAlO_2 + H_2O$. Some metallic oxides having an oxide–oxide bond are known as peroxides. In peroxides, oxygen atoms are present in –1 oxidation state. Some common examples of metallic peroxides are Na_2O_2, K_2O_2, MgO_2, CaO_2, Cu_2O_2, ZnO_2, and so on. Transition metal compounds have electronic structures similar to the noble Pt metal with interstitial phase or interstitial compounds. Transition metal oxides (TMOs) are one of the most important classes of metal oxides or solids because they exhibit a variety of structures and properties. In the periodic table, d block elements are placed in groups 3–12. These are all metallic in nature and also known as transition metals (TMs). These TMs have high electrical conductivity because their valence shell electrons are loosely bounded in d orbital. Like other metals, TMs are also have low ionization potential and variable oxidation states, and so they can form many types of compounds. The nature of metal–oxygen bonding can vary from ionic to highly covalent or metallic. The TMO has a unique nature due to its outer d electrons. It also has extraordinary electronic and magnetic properties. TMO has oxide with metallic properties (e.g., RuO_2, ReO_3, and $LaNiO_3$) at the one end of the range and oxide with insulating properties (e.g., $BaTiO_3$)

at the other end. There are also some oxides that travel across both regimes with the change of temperature, pressure, or composition (e.g., V_2O_3, $La_{1-x}Sr_xVO_3$). Interesting electronic properties also arise due to their charge density waves (e.g., $K_{0.3}MoO_3$), charge ordering (e.g., Fe_3O_4), and defect ordering (e.g., $Ca_2Mn_2O_5$, $Ca_2Fe_2O_5$) [13, 14]. Recently transition metal oxides and their composites have received a lot of attention in DSSCs because of their advantages. Metal oxide composites are used as a counter electrode in DSSC applications. Transition metal compounds are the best alternative materials for Pt-free CE. In transition metal compounds, metal oxides are using very less as compared to metal carbides, nitrides, and chalcogenides for CE preparation in DSSCs. However, some research groups are trying to replace Pt by MO materials in CE. Moreover, MO CEs are showing PCE nearly equal to or better than Pt CE [15, 16]. Different types of MOs as CE in DSSCs are shown in Fig. 2.2.

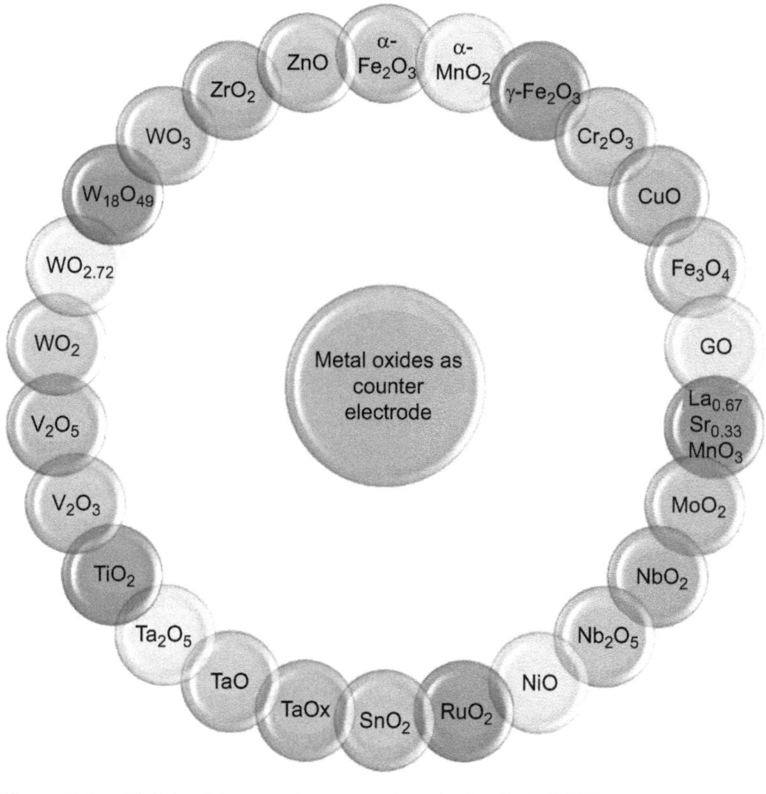

Figure 2.2 Metal oxides used as a counter electrode in DSSCs.

2.5 Preparation of Metal Oxides

Preparation of metal oxides is the simplest way and also an economical method. They can be prepared by a number of methods, namely, wet chemical synthesis [10, 17–20], sol–gel [21], in situ chemical synthesis [22], thermal decomposition [23], vapor deposition [24], high-temperature solid state [25], hydrothermal [26], solvothermal [17, 27], and pyrolysis [28].

2.5.1 Sol–Gel Method

Sol–gel is a most common method to prepare metal oxide materials. It converts the monomers into colloidal solution (sol), which acts as a precursor for integrating the network (or gel) of discrete particles or network polymers. Control of the size distribution and stability of the metal oxides can be done by varying experimental parameters, including dopant variation, heat treatment and surfactants, including inverted micelles, polymer matrix, copolymer, polymer blends, porous glasses, and ex situ particle capping. However, the basic chemistry of sol–gel method is complex due to a variety of reaction parameters and different reactivities of the network modifying and network forming component [29, 30].

2.5.2 Hydrothermal and Solvothermal Methods

Hydrothermal method is generally defined as the crystal growth or crystal synthesis at high pressure and high temperature in aqueous solutions from substances that are insoluble in normal pressure (<1 atm) and normal temperature (<100°C). The crystal growth is carried out in an autoclave, and inside the autoclave nutrient is added along with water. In growth vessel, a temperature gradient is maintained at one end, which is the hot side. It dissolves the nutrient and the cooler end causes the seed to get additional growth. Hydrothermal method is more suitable for growth of large crystal with better quality while maintaining or controlling their composition. The disadvantages of the hydrothermal method include the need of expensive autoclave, requirement of seeds of fair size and good quality, and observing the growth is complex [31]. The solvothermal method is the same as the hydrothermal method with small variations. If aqueous or water media is used, it is hydrothermal, whereas in solvothermal [27], solvent is used as base. In this method, solvent under moderate to high temperature (typically between 100°C and 1000°C) and pressure (typically between 1 atm and 10,000 atm) facilitates the interaction of precursors during synthesis. By controlling various parameters, including surfactant type, reaction temperature, reaction time, precursor type, and reaction time, the size, shape, and crystallinity of the material can be altered.

2.5.3 Vapor Deposited Method

Vapor deposition method is used to deposit thin films on substrates or above the layer. In this method, the required material is vaporized and then condensed through condensation, conversion, or chemical reaction to form solid thin films or materials. This method is used to modify or correct the mechanical, electrical, optical, thermal, and corrosion resistance properties. Vapor deposition usually takes place in a vacuum chamber or vacuum atmosphere. It is of two types: (i) physical vapor deposition (PVD) and (ii) chemical vapor deposition (CVD). PVD is defined as the flux of atoms or clusters of atoms condensed at the surface of the substrate in a vacuum medium. The condensed atoms may be neutral or ionized. PVD coating takes place with or without additional chemical reactions. In CVD, the source material is not pure. It is mixed with a volatile precursor that acts as a carrier. The mixture is deposited on the substrate. When the mixture sticks onto the substrate, the precursor eventually decomposes to create a layer of the source material on the substrate. Then the by-products are removed from the chamber through gas flow. PVD and CVD processes are different, but the goal is the same. PVD is an expensive technique, but it has some advantages, such as control of the atomic level chemical composition and film thickness. Moreover, PVD does not require toxic precursor and toxic chemical by-products [24].

2.5.4 Thermal Decomposition Method

Thermal decomposition is a process of chemical decomposition that is caused by heat. The substance decomposition temperature is required to decompose the substance from single substance into two or more simple substances. It is an endothermic reaction and heat is required to break the bond [23, 32].

2.6 Metal Oxide Composites as Counter Electrode

The composite materials are made up of two or more materials. They have unique properties because of their composites [32]. For example, MOs have high electrocatalytic activity due to a large surface area with active sites, but they have poor conductivity, and so the best conductivity materials, such as conducting polymers and SWCNTs (single-walled carbon nanotubes), are combined with MOs to prepare MO composite to improve the PCE of the MO composite CE. C.-H. Tsai et al. [33] synthesized CuO and its composites (CuO/graphene) and used them as CE for DSSC applications. CuO/graphene CE has the PCE (3.40%) higher than the CuO CE PCE (2.73%). The electrochemical catalytic properties of CE were studied through CV graph (Fig. 2.3a) by observing the peak values of the reduction reaction. In CV graph, +0.6 V to +1.2 V corresponds to the oxidation reaction, whereas −0.6 V to −1.2 V corresponds to the reduction reaction. The CuO reduction current peak value is 2.41 mA/cm^2, but the CuO/graphene reduction current peak value is 4.12 mA/cm^2. As a result, the graphene enhances the electrochemical catalytic activity due to superior oxidation and reduction peak values. This composite CE PCE (3.40%) is closer to the conventional Pt CE (6.36%). J. Xia et al. [34] prepared V_2O_5/Al composite for the solid-state DSSC application. This composite indicates above 2% PCE, which is comparable to novel Ag CE. C.-S. Chou et al. [35] prepared p-type NiO and its composite (Pt(E)/NiO) CE for DSSCs. The prepared CE with different thicknesses gives variation in PCE, but 12 μm thickness composites deliver PCE of 3.51%, which is better than the conventional Pt (3.00%) (Fig. 2.3c). The presence of NiO film in Pt(E)/NiO enhances the current density of redox process. In the oxidation process, the current density of Pt(E)/NiO (7.2 mA/cm^2) is higher than the Pt(E) (6.1 mA/cm^2). In the reduction process, the current density of Pt(E)/NiO (–5.2 mA/cm^2) is lower than the Pt(E) (–3.9 mA/cm^2). The active surface area is increased due to the deposition of NiO and this increases the electrocatalytic activity. M. N. Mustafa et al. [36] prepared poly(3,4-ethylenedioxythiophene)-graphene oxide/titanium dioxide (PEDOT-GO/TiO_2) composite-based CE for the DSSC application. PEDOT-GO/TiO_2 CE possesses a low R_{ct} and has a high electrocatalytic activity. The EIS (electrochemical impedance spectroscopy) Nyquist plot (Fig. 2.3e) indicates that the PEDOT-GO/TiO_2 composites have lower R_s (22.2 Ω) value than that of PEDOT-GO (30.6 Ω) and platinum (175.8 Ω), indicating the better electrical conductivity. It has also smaller R_{ct} (9 Ω) value, indicating the good electrocatalytic activity. A good enhancement in electrochemical conductivity and electrical conductivity is due to the high surface area of GO, high conductivity of PEDOT, and high active area of TiO_2. PEDOT-GO/TiO_2 composite CE has a better PCE (1.166%) than Pt CE PCE (0.727%). W. Cui [37] prepared tungsten trioxide (WO_3) loaded on activated charcoal powder (ACP) (WO_3@ACP) for DSSC. The CV results revealed that ACP, WO_3, and WO_3@ACP electrodes have two pairs of redox peaks corresponding to $I^{3-} + 2e^- \rightleftharpoons 3I^-$ and $3I^2 + 2e^- \rightleftharpoons 2I^{3-}$. The CV results (Fig. 2.3b) indicate that the greater the peak current, the smaller the peak potential difference. The good electrocatalytic behavior of WO_3 CE has the current density reduction peak higher than that of ACP. The peak to peak separation (ΔE_p) of left redox peak for WO_3@ACP (0.11 V) is smaller than the WO_3 (0.45 V). Finally, CV results indicate that the WO_3@ACP composite CE has better electrocatalytic activity. WO_3@ACP composites have higher PCE (5.04%) than that of WO_3 CE (1.61%) and ACP CE (2.10%). X. Chen et al. [38] prepared a novel Pt–SnO_2 nanocomposite high-efficient CE for DSSC. The EIS Nyquist plot (Fig. 2.3f) shows the R_s, R_{ct}, and constant phase element (CPE). The R_s value of Pt–SnO_2 CE (21.46 Ω) is smaller than that of Pt CE (21.95 Ω). This smaller R_s value indicates the excellent electrical conductivity and electron transfer kinetics compared to those of Pt CE. The R_{ct} is the measure of electron exchange at CE and electrolyte and inversely triiodide reduction of CE. The R_{ct} is slightly lower than the Pt CE, which indicates the better catalytic activity (Fig. 2.3d). The high capacitance of Pt–SnO_2 indicates the high surface area than that of the Pt CE, which enhances the Pt–SnO_2 CE PCE (8.83%) compared to Pt CE PCE (8.04%). H. Wang et al. [39] prepared NiO and NiO composite CE for DSSCs. The NiO CE has negligible PCE (0.28%) due to low electrical conductivity; the electrical conductivity of the CE is increased by adding the PEDOT:PSS conducting polymer (Fig. 2.3g). PEDOT:PSS alone gives 4.6% PCE, but the NiO/PEDOT:PSS (48/1) composites gives 7.58% PCE when compared to Pt CE (8.7%). D. Song et al. [40] prepared inorganic/organic (SiO_2/PEDOT–PSS) highly transparent and efficient CE for bifacial DSSCs. This composite CE provides better transmission in long-wavelength range due to PEDOT–PSS content. The R_s value of SiO_2/PEDOT–PSS CE (14.5 Ω) is high compared to that of PEDOT–PSS CE (10.1 Ω) due to the dielectric property of SiO_2. The R_{ct} value of SiO_2/PEDOT–PSS CE (1.4 Ω) is small compared to that of PEDOT–PSS CE (6.6 Ω), which indicates the better catalytic for reduction of triiodide. The PCE of SiO_2/PEDOT–PSS CE (5.66 %) is small compared to Pt CE (6.38 %) (Fig. 3.3i). M. Chen et al. [41] prepared Co_3O_4-MC-*x* composite CE, where *x* represents the added amount of cobalt nitrate from the initial stage and it is labelled as Co_3O_4-MC-1, Co_3O_4-MC-2, and Co_3O_4-MC-4. Here, Co_3O_4 acts as a catalyst and the catalytic activity is increased due to the loading of Co_2N, which is shown (Fig. 2.3h) in EIS studies. The conductivity is increased due to graphitic mesoporous carbon. The Co_3O_4-MC-4 CE PCE (5.08%) is higher than the Pt CE PCE (4.88%). M. R. Al-bahrani et al. [42] prepared RGO/MWCNTs/NiO CE to enhance the electrocatalytic activity of CE in DSSC. The NiO acts as an excellent electrocatalytic activity, while RGO acts as an electrical conductivity material. The MWCNTs (multi-walled carbon nanotubes) not only act to prevent the aggregation of NiO/RGO, but also increase the electron transport property of CE due to their special conductivity property. The RGO/MWCNTs/NiO CE PCE is higher compared to NiO (2.71%), RGO (6.77%), and NiO/RGO (7.63%). S.-S. Kim et al. [43] prepared Pt–NiO and Pt–TiO_2 biphase CEs. These CEs, consist of nanosized Pt polycrystalline phase mixed with NiO or TiO_2, for DSSC were fabricated by co-sputtering method. The NiO combined with Pt not only increases the electrocatalytic activity, but also increases both the active surface area and the adhesion. The Pt–NiO (3.62%) and Pt–TiO_2 (4.21%) CE PCE is higher than the Pt CE (2.70%). M. R. Al-bahrani et al. [44] prepared NiO-NF/MWCNT nanocomposite CE. While NiO-NF (nanofilament) has high electrocatalytic activity, MWCNT has excellent conductivity due to its special property. This NiO-NF/MWCNT CE (7.63%) provides higher PCE than the Pt CE (6.72%).

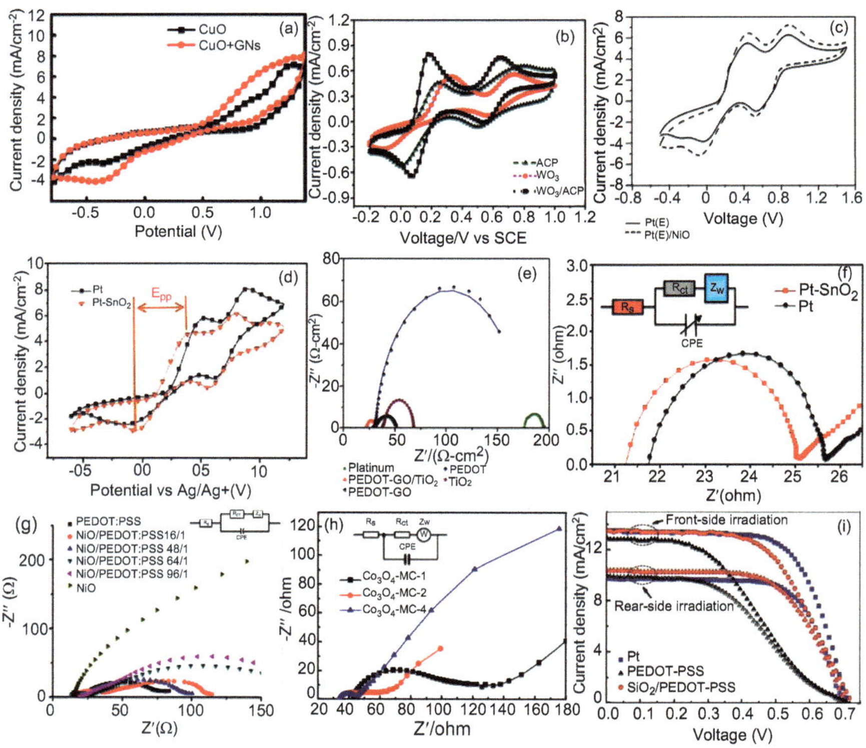

Figure 2.3 CV plots of (a) CuO and CuO+GNs [33]; (b) WO_3 and WO_3/ACP [37]; (c) Pt (E)/NiO [35]; and (d) Pt-SnO_2, reprinted from Ref. [38], with permission from the Royal Society of Chemistry. EIS Nyquist plots of (e) PEDOT-GO/TiO_2 [36]; (f) Pt-SnO_2, reprinted from Ref. [38], with permission from the Royal Society of Chemistry; (g) NiO and NiO/PEDOT:PSS, reprinted from Ref. [39], with permission from Springer Nature; and (h) Co_3O_4-MC-*x*, reprinted from ref. [41], with permission from Elsevier. (i) *J–V* curves of SiO_2/PEDOT-PSS, reprinted from Ref. [40], with permission from American Chemical Society.

2.7 Effects of Phase Structures/Bandgap/Morphology on Metal Oxides and Their Composite CEs

Phase structures and morphology are playing a very important role in MOs for improving the catalytic properties of MO CE in DSSCs. Phase structures are confirmed through X-ray diffraction (XRD) spectroscopy and transmission electron microscopy (TEM) analysis, and morphology is identified through SEM (scanning electron microscopy) and TEM. Morphology has attracted attention due to its enhanced unique property, which makes it suitable for various applications. Therefore, much attention has been given to synthesize and characterize different morphologies, for example, 1D structures such as rods, wires, ribbons, whiskers, fibers, belts, needles, tubes, hooks, cables, tips, helices, and zigzags; two-dimensional (2D) structures such as plates, prisms, disks, sheets, and walls; and three-dimensional (3D) structures such as pillers, cones, balls, coils, and flowers. X. Lin et al. [45] synthesized niobium oxides (Nb_2O_5 and NbO_2) using the sol–gel method, followed by different annealing temperatures (500°C, 800°C, 1000°C, 1100°C) for 3 h under nitrogen atmosphere. It shows different phase structures (Fig. 2.4a), such as H-Nb_2O_5 (hexagonal at 500°C), O-Nb_2O_5 (orthorhombic at 800°C), M-Nb_2O_5 (monoclinic at 1000°C), and T-NbO_2 (tetragonal at 1100°C). From these phase structures, tetragonal niobium oxide (T-NbO_2) CE PCE (7.88%) is better than Pt CE PCE (7.65%) (Fig. 2.4b). The remaining phase structures of niobium oxide CE PCEs are lower than the T-NbO_2 CE PCE and Pt CE PCE. The calculated PCE shows that the tetragonal niobium oxide CE acts as a better catalyst than the remaining phase structures of niobium oxides. Figure 2.4c shows that the CV plot have two pairs of oxidation/reduction peaks, the negative pair is indicating the reduction and oxidation of I^-/I_3^-, whereas positive pair is indicating the reduction and oxidation of I_2/I_3^-. The profile and locations of the two pairs of redox peaks for NbO_2 are similarly matching with those of the Pt CE peaks. It signifies better catalyst activity of NbO_2 compared to other niobium oxides. Figure 2.4d shows the Nyquist plot of niobium oxides and Pt. It indicates that NbO_2 and Pt have lower Z_N than other niobium oxides because NbO_2 has a honeycomb structure with a number of porus channels in morphology. These porus channels make the redox couple smoothly diffuse in the electrolyte. So NbO_2 gives a higher catalytic activity than other niobium oxides.

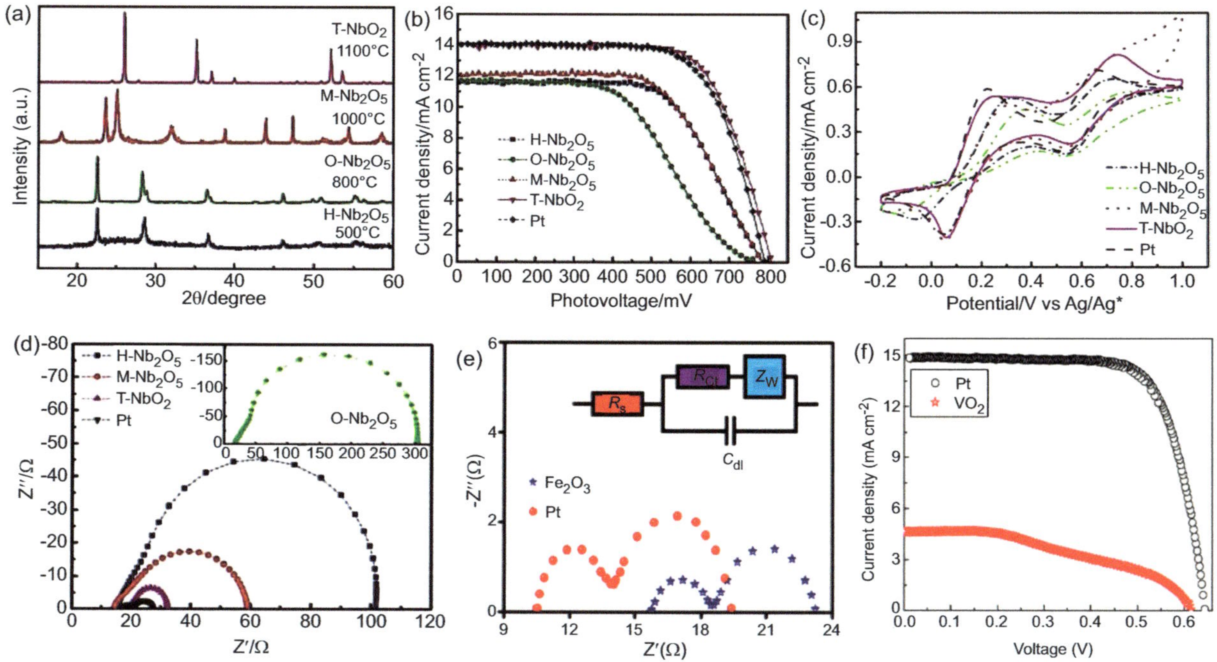

Figure 2.4 (a) XRD patterns of H, O, M-Nb_2O_5, and T-NbO_2, reprinted from Ref. [45], with permission from the Royal Society of Chemistry. (b) *I–V* curves of the DSSCs based on the three Nb_2O_5 CEs, an NbO_2 CE, and a Pt CE, reprinted from Ref. [45], with permission from the Royal Society of Chemistry. (c) Cyclic voltammograms of for five (three Nb_2O_5, an NbO_2, and a Pt) electrodes, reprinted from Ref. [45], with permission from the Royal Society of Chemistry. Nyquist plots of the symmetrical cells based on (d) various phase structures of niobium oxide, reprinted from Ref. [45], with permission from the Royal Society of Chemistry, and (e) Fe_2O_3 CE and Pt CE, reprinted from Ref. [46], with permission from Springer Nature. (f) *J–V* characteristics of VO_2 CE, reprinted from Ref. [47], Copyright 2016, with permission from Elsevier.

2.7.1 Nanoparticles

Y. Hou et al. [46] prepared rational screening low-cost CE in DSSC using α-Fe_2O_3 nanoparticles (Fig. 2.5d). The EIS Nyquist plot (Fig. 2.4e) shows two semicircles in the high-frequency region (left) and low-frequency region (right). The high-frequency region semicircles arise due to R_{ct}. The α-Fe_2O_3 CE has high R_s and small R_{ct} in contrast to Pt CE (7.32%). The α-Fe_2O_3 CE PCE is smaller compared to Pt CE. The α-Fe_2O_3 CE has larger R_s and smaller R_{ct} compared to Pt CE. The PCE (6.96%) of α-Fe_2O_3 CE is very close to Pt CE PCE (7.32%). G. R. Mutta et al. [47] prepared sphere-like VO_2(M1) polymorph (Fig. 2.5b) by the single-step hydrothermal method. This VO_2(M1) polymorph is used as economically viable CE in DSSC. The V_{oc} of both Pt and VO_2(M1) is similar, but the J_{sc} value of VO_2(M1) CE is lesser than that of Pt CE. The PCE of VO_2(M1) CE (1.25%) is lesser than the Pt CE PCE (7.08%) (Fig. 2.4f). M. W. Khan et al. [48] prepared hollow mesosphere shaped SnO_2 nanoparticles (Fig. 2.5a) and its composites SnO_2@SWCNTs@rGO. The SWCNT was dispersed homogeneously on rGO in which SWCNT increases the electrical conductivity and rGO preserves the physical and electrical properties. Therefore, the electrochemical (Fig. 2.5e) and photocatalytic activities were increased. As a result, the PCE of SnO_2@SWCNTs@rGO CE (6.1%) was equivalent to Pt CE (6.2%). P. Vijayakumar et al. [49] prepared a novel CE using WO_3 nanoparticles with large surface (Fig. 2.5c). The BET surface area analysis (Fig. 2.5f) confirms that WO_3 nanoparticles have a large surface area (240.56 m^2/g) and pore volume (0.1157 cc/g). The large surface area indicates that WO_3 nanoparticles have a large number of active sites. The WO_3 nanoparticle CE PCE (0.98%) is small compared to Pt CE (1.72%). Y. Yang et al. [50] prepared mesoscopic indium tin oxide counter electrode (ITO) for fully printable transparent solid-state dye-sensitized solar cells (SS-DSSCs). The ITO nanoparticle CE at 21 μm thickness provides 1.73% PCE in front-side irradiation and 1.06% PCE in rear-side irradiation. Y.-F. Chiang et al. [51] also prepared ITO CE for SS-DSSCs. It gave PCE of 1.96% under front-side irradiation and PCE of 1.50% under back-side irradiation. W. Ahmad et al. [52] prepared P-type nickel oxide acetylene black nanoparticle composite (NiO/AB) CE for DSSC. NiO acts as a better catalyst due to its valence band structure which collects holes from I_3^-/I^- and also have strong interaction with electrolytes of iodine-based solution. The NiO has poor conductivity, and so an electrically conductive component is added to increase its electrical conductivity. The NiO/AB PCE (7.75 ± 0.3) is higher than the Pt CE (6.96 ± 0.3).

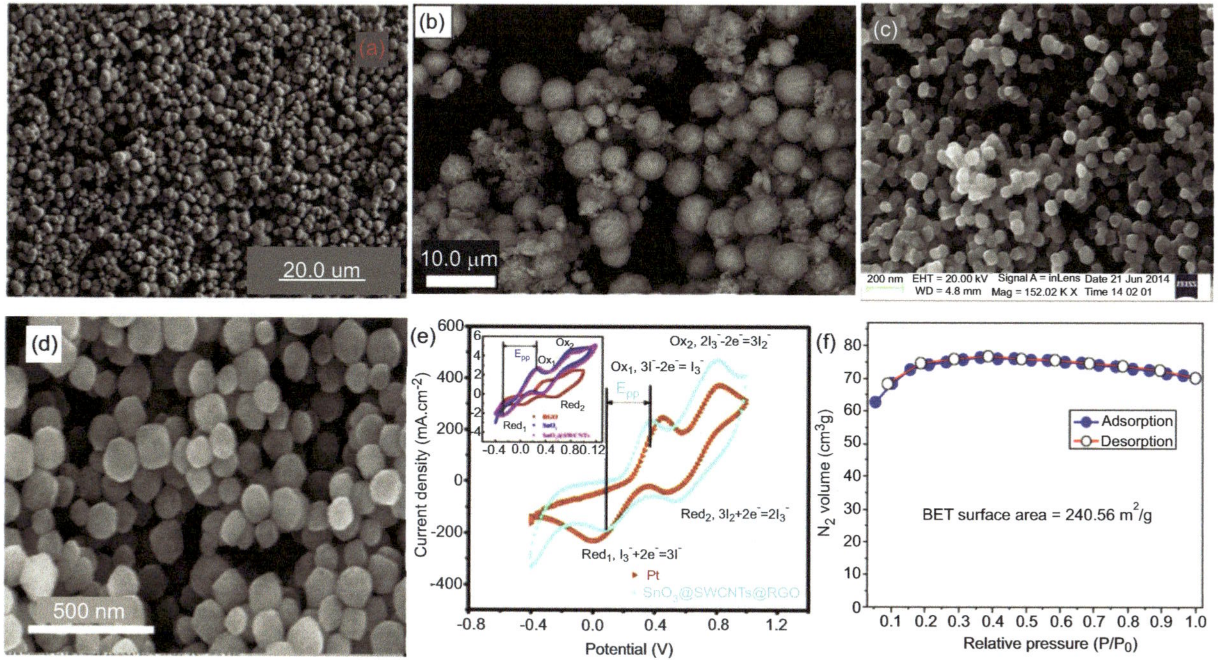

Figure 2.5 SEM images of (a) homogeneous formation of hollow-shell mesospheres of crystalline SnO_2 nanoparticles, reprinted from Ref. [48], with permission from Elsevier; (b) VO_2 nanoparticles, reprinted from Ref. [47], with permission from Elsevier; (c) WO_3 nanoparticles, reprinted from Ref. [49], with permission from Springer Nature; and (d) α-Fe_2O_3 nanoparticles, reprinted from Ref. [46], with permission from Springer Nature. (e) CV plots of SnO_2@SWCNTs@rGO CE and Pt CE, reprinted from Ref. [48], with permission from Elsevier. (f) N_2 adsorption–desorption isotherms for tungsten oxide (WO_3) using the BET method, reprinted from Ref. [49], with permission from Springer Nature.

2.7.2 Nanorods

H. Wang et al. [53] prepared ZnO composites using a conducting polymer [poly(3,4-ethylenedioxythiophene):polystyrene sulfonate (PEDOT:PSS)]. ZnO acts as a catalyst but is a poor conductor, and so conducting polymer is combined with ZnO to increase the electrical conductivity to act as an efficient CE in DSSCs. The ZnO composites as CE were synthesized by varying ZnO and conducting polymers with a weight ratio of 4:1, 8:1, 16:1, 32:1, 48:1, or 64:1. The different variations of ZnO composites give different values of PCE, but 32:1 (ZnO/polymer) ratio shows the highest PCE (8.17%) (Fig. 2.6e). B.-G. Zhai et al. [54] prepared ZnO nanorods decorated graphene/ZnO nanoparticle composite CE (Fig. 2.6c,d). ZnO nanoparticles with large surface area ZnO nanorods provide excellent electrocatalytic for the reduction of triiodide to iodide. ZnO nanoparticles prevent the aggregation of graphene nanosheets to maintain excellent conductivity of the graphene. Figure 2.6a,b shows that the prepared ZnO composite CE PCE (0.17%) was better than the conventional Pt CE (0.15%). H. Zhou et al. [55] synthesized $WO_{2.72}$ nanorod bundles (NRB) via the solvothermal method. After that this $WO_{2.72}$ nanorod was annealed at 500°C in the air for 30 min to form WO_3 prisms. Both materials show monoclinic crystal structures (Fig. 2.6f). Moreover, the XRD pattern of $WO_{2.72}$ indicates both narrow peaks (010) and (020). This shows that the NRB growth is highly anisotropic along the [010] direction, and it is also confirmed through TEM. The XRD patterns (Fig. 2.6f) show that both materials have the same phase structure with very little difference. This small difference gives the various PCEs. $WO_{2.72}$ NRB shows PCE (8.03%) very close the Pt CE PCE (8.08%) using iodine–triiodide electrolyte, but WO_3 prisms show very low PCE (4.60%) as compared to Pt CE PCE (Fig. 2.6g). Both materials ($WO_{2.72}$ NRB, WO_3 prisms) show high PCE [$WO_{2.72}$ NRB (5.05%), WO_3 prisms (4.17%)] as compared to Pt CE PCE (2.73%) in thiolate electrolyte (Fig. 2.6h). Therefore, the calculated PCE indicates that the catalytic activity of tungsten oxides is also based on the phase structure of materials.

G. R. Mutta et al. [56] prepared V_2O_5 nanorods and used them for CE preparation in DSSC. The performance of the H-V_2O_5 CE was compared with the commercial V_2O_5 (C-V_2O_5). It shows (Fig. 2.6j) different optical bandgaps in both materials, that is, H-V_2O_5

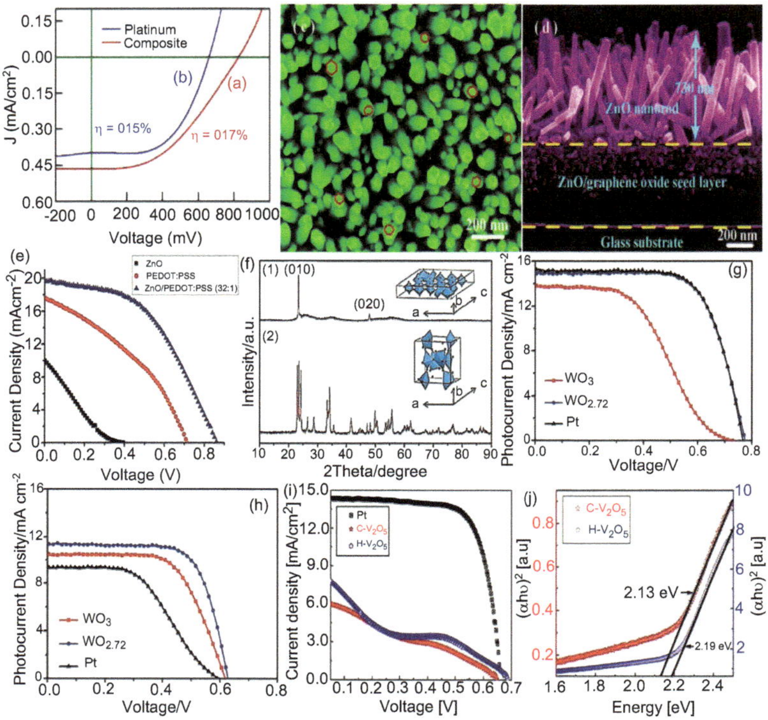

Figure 2.6 *J–V* characteristic curves of DSSCs with counter electrodes: (a) conventional Pt CE, reprinted from Ref. [54], with permission from Taylor & Francis; (b) ZnO nanorods decorated graphene/ZnO composite film as the CE, reprinted from Ref. [54], with permission from Taylor & Francis. Top-view and cross-sectional SEM micrographs of CEs: (c, d) ZnO nanorods decorated graphene/ZnO composite film on FTO glass substrate, reprinted from Ref. [54], with permission from Taylor & Francis. (e) *J–V* curves of DSSCs with CEs of ZnO, PEDOT:PSS, and ZnO/PEDOT:PSS (weight ratio 32:1), reprinted from Ref. [53], with permission from the Royal Society of Chemistry. (f) XRD patterns of (1) as-synthesized $WO_{2.72}$ NRBs and (2) WO_3 prisms; the insets show crystal structures of monoclinic $WO_{2.72}$ and WO_3, reprinted from Ref. [55], with permission from the Royal Society of Chemistry. Photocurrent-voltage (*J–V*) curves of DSSCs based on two different electrolytes containing (g) and (h) organic redox couples, using $WO_{2.72}$ NRBs, WO_3 and Pt as CEs, reprinted from Ref. [55], with permission from the Royal Society of Chemistry. (i) *J–V* curves of the DSSCs fabricated with C-V_2O_5, H-V_2O_5, and Pt CEs, reprinted from Ref. [56], with permission from Materials Research Express. (j) Plot to estimate the direct optical bandgap of C-V_2O_5 and H-V_2O_5 powder materials, reprinted from Ref. [56], with permission from Materials Research Express.

(2.19 eV) and C-V_2O_5 (2.13 eV). The prepared high bandgap material CE shows the PCE (1.6%) higher than the commercially purchased low bandgap material CE PCE (1.2%) (Fig. 2.6i). T. Balkan et al. [57] prepared TiO_2 nanorods and used as one of the materials of TiO_2-PEDOT:PSS nanocomposite electrodes. The nanorods with different weight percentages are mixed with PEDOT:PSS and labelled as NR0, NR25, NR50, and NR100 (Fig. 2.7d). The catalytic activity of TiO_2-PEDOT:PSS nanocomposite CE is significantly increased due to TiO_2 nanorods. Although NR0 has the highest surface area or electroactive area, NR100 exhibits almost the same catalytic activity for the reduction of I_3^-. Moreover, it indicates that R_{ct} is mostly dependent on TiO_2 contents. The increase in TiO_2 content in composites will decrease the R_{ct}. Figure 2.7e shows the Nyquist plot of PEDOT:PSS and TiO_2-PEDOT:PSS composites. It indicates that NR100 has low sheet resistance and prominent conductivity. Figure 2.7g shows CV curves of all samples. In these samples, except PEDOT:PSS curves, the remaining sample curves have two pairs of redox peaks. The left reduction peaks show the reduction of triiodide ($I_3^- + 2e \rightarrow 3I^-$) and the right peaks correspond to $3I_2 + 2e \rightarrow 2\ I_3^-$. In DSSC, left reduction peaks are important to show the catalytic activity. In different weight percentages of TiO_2, NR100 shows the best electroactive area and the best electron transport in TiO_2-PEDOT:PSS nanocomposite CEs. S. Yun et al. [58] synthesized tantalum oxides (TaO and Ta_2O_5) using the sol–gel method, followed by sintering in the temperature range of 600–1100°C for 3 h under N_2. The different sintering temperatures give various structures (Fig. 2.7h) and morphologies (Fig. 2.7f). The synthesized powder sintered below 800°C gave hexagonal tantalum oxide structures, and it also gave the morphology of hexagonal pillar-shaped TaO. The synthesized sample sintered above 1000°C gave orthorhombic pentoxide phase structures as well as the aggregated nanorod morphology. The CV curves (Fig. 2.7i) of TaO and Ta_2O_5 CE show two pairs of redox peaks, which indicate the high catalytic activity of I_3^- reduction. The TaO CE shows the two pairs of redox peaks similar to the Pt CE, which indicates that TaO also behaves like Pt CE in regard to the electrocatalytic activity of I_3^- reduction. The tantalum oxide CE PCE (6.48%) is very close to the conventional Pt CE (7.16%), but tantalum pentoxide CE PCE (4.08%) is low compared to TaO and Pt CEs.

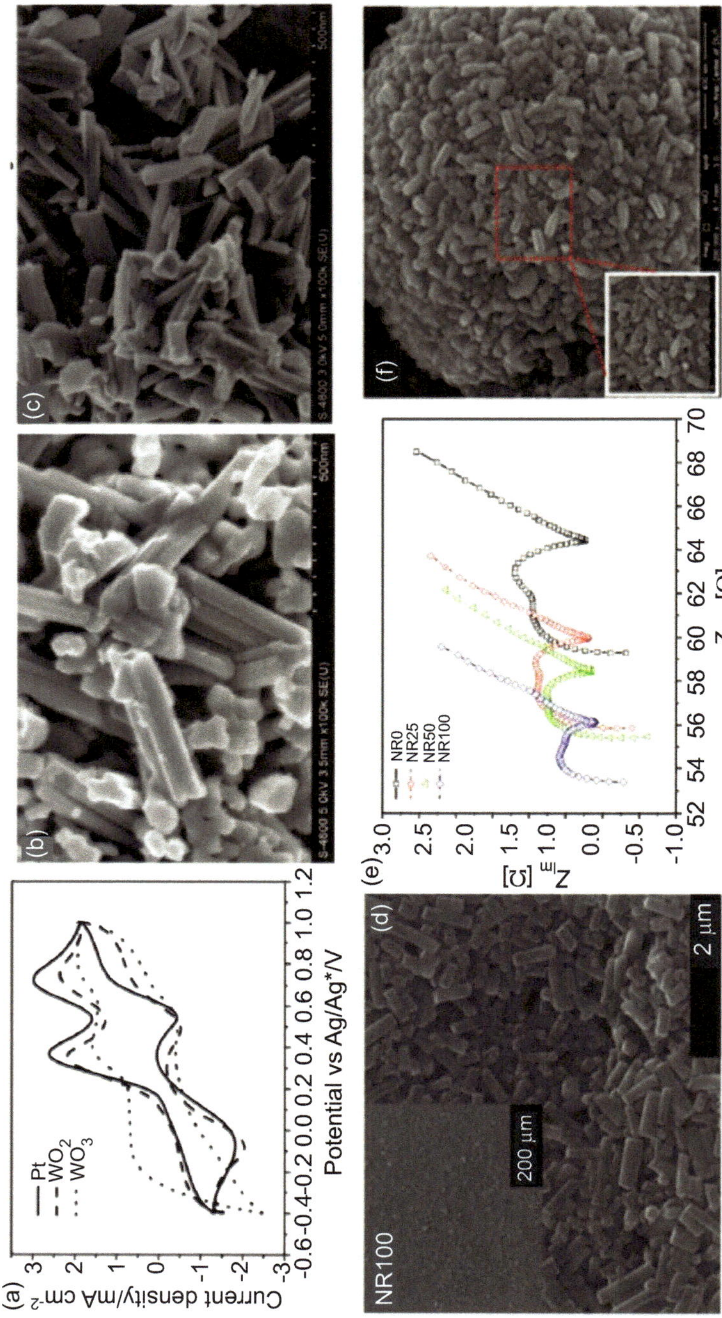
(a)
Pt
WO2
WO3
Current density/mA cm-2
Potential vs Ag/Ag+/V
(b)
(c)
(d)
NR100
200 μm
2 μm
(e)
NR0
NR25
NR50
NR100
Z_Im [Ω]
Z_Re [Ω]
(f)

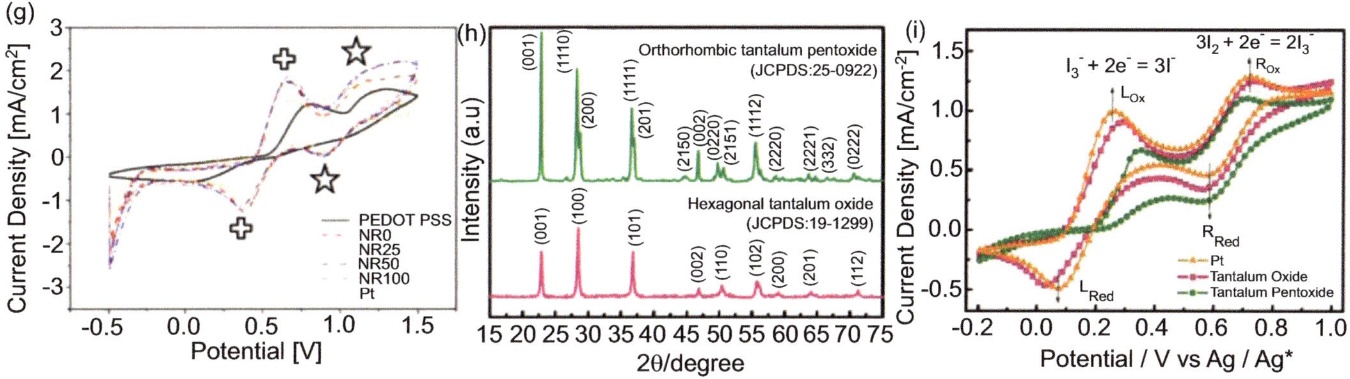

Figure 2.7 (a) Cyclic voltammograms of triiodide/iodide redox couple for WO_2, WO_3, and Pt electrodes, reprinted from Ref. [59], with permission from the Royal Society of Chemistry. (b, c) SEM images of WO_2 nanorods, reprinted from Ref. [60], with permission from American Chemical Society, and from Ref. [59], with permission from Royal Society of Chemistry. (d) SEM images of TiO_2-PEDOT:PSS composite films on FTO substrates NR100 [57]. (e) Nyquist plots of NR0, NR25, NR50, and NR100 composite electrodes [57]. (f) SEM images of Ta_2O_5, reprinted from Ref. [58], with permission from Elsevier. (g) CV curves of Pt, PEDOT: PSS, and TiO_2-PEDOT:PSS composite electrodes according to redox couple. Symbol related to reduction reaction and symbol related to [57]. (h) XRD patterns of TaO and Ta_2O_5 materials, reprinted from Ref. [58], with permission from Elsevier. (i) CVs of Pt, TaO, and Ta_2O_5 electrodes in a triiodide/iodide redox couple. The scan rate is 0.01 V/s. Pt electrode is used as CE and Ag/Ag^+ works as reference electrode, reprinted from Ref. [58], with permission from Elsevier.

M. Wu et al. [59] prepared a novel catalyst of WO_2 nanorod (Fig. 2.7c) CE in DSSC. The CV curves (Fig. 2.7a) of WO_2 provide two pairs of redox peaks, the cathodic peak at 0.491 V according to $I_3^- + 2e \rightarrow 3I^-$ reaction and the peak at −0.008 V according to $3I_2 + 2e \rightarrow 2I_3^-$ reaction. The peak profiles of WO_2 are similar to those of Pt CE, which indicates the efficient electrocatalytic activity of WO_2 because of reduction of triiodide to iodide. The WO_2 CE PCE (7.25%) is close to the Pt CE PCE (7.57%). M. Wu et al. [60] also prepared WO_2 nanorods (Fig. 2.7b) and WO_2-MC CE to increase the electrocatalytic activity. The mesoporous carbon (MC) serves as an electrical conductor and WO_2 serves as a catalyst. So both can be imbedded into single material WO_2-MC composites, which provide efficient electrical conductivity and act as catalyst. The CV curves (Fig. 2.8e) of all samples have two pairs of redox peaks, which show that the MC reduction current density is high, whereas WO_2 shows low reversibility of triiodide/iodide redox reaction. The WO_2-MC CE also shows two pairs of redox peaks, but has higher current density and reversibility compared to Pt CE. The WO_2-MC CE shows the PCE (7.76%) higher than the Pt CE PCE (7.55%).

K. Ahmad et al. [61] prepared hydrothermally grown α-MnO_2 nanorods (Fig. 2.8c) and used as CE. The CV curves (Fig. 2.8a) of α-MnO_2 nanorods CE show two pairs of redox peaks in which the negative pair indicates the oxidation–reduction reaction $I_3^- + 2e \rightarrow 3I^-$ and the positive pair indicate the reduction–oxidation reaction $3I_2 + 2e \rightarrow 2I_3^-$. The α-MnO_2 nanorods CE gives better electrocatalytic activity as compared to Pt CE. M. Wu et al. [62] prepared WO_2 nanorods (Fig. 2.8d) and WO_3/WO_2 nanorods (Fig. 2.8b) CE through the one-step chemical synthesis method. In this method, ethanol is used as an oxygen source, which increases the electrochemical catalytic activity to reduce triiodide to iodide. The CV curves (Fig. 2.8f) show two pairs of redox couple in which WO_2 (13.61 mA/cm^2) shows a larger current density than WO_3/WO_2 CE (12.17 mA/cm^2). As a result, it shows the higher catalytic activity than others. The WO_2 CE PCE (6.88%) is very close to the Pt CE (7.65%).

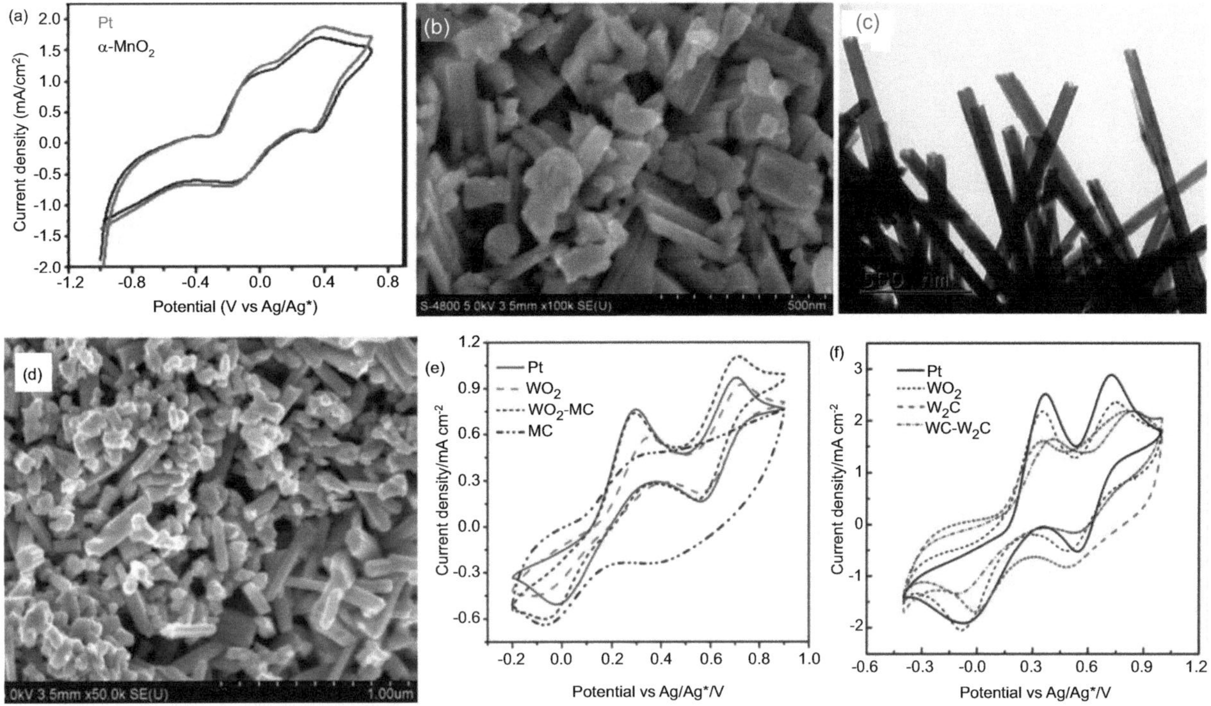

Figure 2.8 (a) CV of the triiodide/iodide redox couple for a MnO_2-based counter electrode (black) and Pt (red), reprinted from Ref. [61], with permission from Elsevier. SEM images of (b) WO_3/WO_2 and (d) WO_2, reprinted from Ref. [62], with permission from the Royal Society of Chemistry. (c) TEM images of a-MnO_2 nanorods, reprinted from Ref. [61], with permission from Elsevier. (e) Cyclic voltammograms of the Pt, WO_2, WO_2-MC, and MC electrodes, reprinted from Ref. [60], with permission from American Chemical Society. (f) Cyclic voltammograms of the as-prepared WC/W_2C, W_2C, WO_2, and Pt electrodes, reprinted from Ref. [62], with permission from the Royal Society of Chemistry.

2.7.3 Nanowires

S. Sui et al. [63] prepared $W_{18}O_{49}$ nanowire (NW)-reduced graphite oxide ($W_{18}O_{49}$-rGO) nanocomposite for DSSC. The perfect ratio of $W_{18}O_{49}$ NW and rGO content gives better morphology and better catalytic activity. If the $W_{18}O_{49}$ NW mass content is decreased, $W_{18}O_{49}$ NW length becomes shorter, and further decrease in the mass content of $W_{18}O_{49}$ NW (61%) reduces the aggregation. Moreover, decrease in the mass content of $W_{18}O_{49}$ NW (33%) gives good dispersion on the surface of rGO with a relatively uniform degree. In $W_{18}O_{49}$-rGO nanocomposites, 33% $W_{18}O_{49}$ is controlled properly and used as a counter electrode in DSSC with PCE of 7.23%, which is closer to Pt (7.39%). This high efficiency is due to the intrinsic electrocatalytic activity of $W_{18}O_{49}$ and excellent electron transfer characteristics of rGO. The principle investigation indicates the good catalytic activity with the help of adsorption energy between I_3^- and $W_{18}O_{49}$ NW. C.-H. Tsai et al. [64] prepared Cu_2O nanostructures sintered at different temperatures (400°C, 500°C, 600°C, and 700°C) under nitrogen atmosphere, where sintering at 400°C gives nanowire structures. This Cu_2O NW CE has smaller PCE (0.43 ± 0.10) compared to other different sintered temperatures [500°C (3.11 ± 0.25), 600°C (3.62 ± 0.31), and 700°C (0.88 ± 0.16)].

2.7.4 Nanofibers

H. W. Zheng et al. [65] prepared $Bi_5FeTi_3O_{15}$ (BFTO) nanofiber/graphene (Gr) nanocomposites (BGr) as CE in DSSCs. BFTO is of perovskite family, which has high chemical stability, is non-toxic, and shows direct bandgap. BFTO nanofibers are shown in Fig. 2.9a,b. BGr CEs are prepared with different weight percentages (wt%) of Gr (0, 0.5, 1, 1.5, and 2) labelled as BGr0, BGr0.5, BGr1, BGr1.5, BGr2, respectively. The optical bandgap (Fig. 2.9d) of BGr decreases with increasing content of Gr, which increases the photoabsorption capability and photocarrier generation. The BGr0 and BGr0.5 composites have a large R_{ct} value, and because of this they have poor catalytic activity. BGr2 has a low R_{ct} value, implying excellent catalytic activity and electric conductivity comparable to (BGr0) pure BFTO fibers due to the large surface area and conductivity of CE. The PCE of BGr2 (9.56%) is higher than that of BGr0 (0.22%). H. Zhou et al. [66] prepared interlaced $W_{18}O_{49}$ nanofibers (NFs)

(Fig. 2.9c) and nanofiber bundles (NFBs) by a facile and template-free solvothermal method. The EIS Nyquist plots (Fig. 2.9e) of both $W_{18}O_{49}$ NFs and NFBs show two irregular semicircles. While the first semicircle specifies R_{ct} at CE–electrolyte interface, the second semicircle specifies Z_N within the electrolyte. The R_s values of $W_{18}O_{49}$ CE are larger than the Pt CE. The R_{ct} values of both $W_{18}O_{49}$ NFs (0.98 Ω) and NFBs (1.18 Ω) are smaller than the Pt CE (3.30 Ω). Both $W_{18}O_{49}$ NFs and NFBs have good catalytic activity because of their structural advantages, such as abundant reaction sites, through contact with the electrolyte solution, large specific areas, and their interlaced networks. These NFs and NFBs CEs have good electron transport because of their interlaced NF networks.

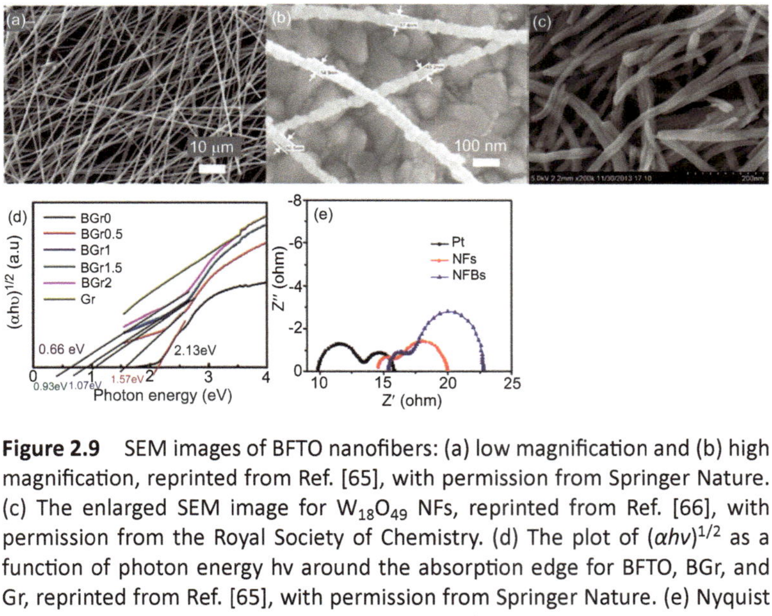

Figure 2.9 SEM images of BFTO nanofibers: (a) low magnification and (b) high magnification, reprinted from Ref. [65], with permission from Springer Nature. (c) The enlarged SEM image for $W_{18}O_{49}$ NFs, reprinted from Ref. [66], with permission from the Royal Society of Chemistry. (d) The plot of $(\alpha hv)^{1/2}$ as a function of photon energy hv around the absorption edge for BFTO, BGr, and Gr, reprinted from Ref. [65], with permission from Springer Nature. (e) Nyquist plots of $W_{18}O_{49}$ NFs and $W_{18}O_{49}$ NFBs, reprinted from Ref. [66], with permission from the Royal Society of Chemistry.

2.7.5 Nanoflowers

L. Wang et al. [67] synthesized iron oxide nanoflowers (Fig. 2.10a) with different phase structures (Fe_3O_4 to α-Fe_2O_3 or γ-Fe_2O_3) through the hydrothermal method, followed by different annealing temperatures. These different phase structures (Fig. 2.10d) of iron oxide are used as CE in DSSCs and they give different PCEs (Fe_3O_4 = 6.80%, α-Fe_2O_3 = 5.59%, γ-Fe_2O_3 = 6.39%), which are close to the PCE of the conventional Pt (sputtered Pt = 7.9%, pyrolytic Pt = 6.89%). In Fig. 2.10e, the CV plot shows two pairs of redox peaks in all the curves. The redox peaks indicate the following reactions: $I_3^- + 2e^- \leftrightarrow 3I^-$ (for lower potential peaks) and $3I_2 + 2e^- \leftrightarrow 2I_3^-$ (for higher potential peaks). The CV plots show the cathodic reduction peak (Red-1) and anodic oxide peak (Ox-1), which correspond to the reduction of I_3^- and oxidation of I^-, respectively. This CV plot indicates small peak to peak separations (E_{pp}) and high cathodic peak current density of Fe_3O_4, specifying the higher electrocatalyst than other iron oxide CEs. Tafel polarization curves (Fig. 2.10f) show that Fe_3O_4 has a larger slope than other iron oxides. It specifies the higher exchange current density (J_0) and limiting current density (J_{lim}). Moreover, the Nyquist plot based on symmetrical cells has two semicircles—the left semicircle specifies charge transfer (R_{ct}) resistance at CE/electrolyte interface and the right semicircle specifies the Nernst diffusion impedance (Z_N) of redox couple in the electrolyte. The R_{ct} value of Fe_3O_4 CE is much lower than that of other iron oxide CEs and very close to that of sputtered Pt. Therefore, Fe_3O_4 CE is more active than other iron oxide CEs for reducing I_3^-. M. Shahpari et al. [68] prepared hematite (α-Fe_2O_3) using the solvothermal method. The flowers (Fig. 2.10b,c) are composed of nanosheets with 70 nm thickness and 1–2 μm width agglomerated to each other. The CV curves of α-Fe_2O_3 CE show that the two reduction peaks are small and overlap each other, which indicate that the reduction activity is carried out slowly. The α-Fe_2O_3 nanoflowers CE gives smaller PCE (3.19 ± 0.27%) compared to Pt PCE (5.29 ± 0.13%).

2.7.6 Honeycomb-Like Structure and Nanotubes

A. T. Yadav et al. [69] synthesized NiO film by the chemical deposition method. The prepared NiO film shows the porous honeycomb-like structure (Fig. 2.11a) with coverage of the substrate surface. This porous honeycomb-like structure gives a high surface to volume ratio, which may increase the dye absorption in DSSC. This porous structure enhances the contact area of the electrolyte, which increases the conduction of electrons. The PCE (0.02%) of NiO film CE is double than the Pt CE PCE (0.01%).

Figure 2.10 The SEM images of flower-like (a) Fe3O4, reprinted from Ref. [67], with permission from the Royal Society of Chemistry; and (b, c) α-Fe_2O_3 - nanoflowers, reprinted from Ref. [68], with permission from Elsevier. (d) XRD patterns of α-Fe_2O_3, Fe_3O_4, and γ-Fe_2O_3, reprinted from Ref. [67], with permission from the Royal Society of Chemistry. (e) Cyclic voltammograms based on iron oxides and Pt CEs in the three-electrode system, reprinted from Ref. [67], with permission from the Royal Society of Chemistry. (f) Tafel curves of iron oxides and Pt CEs, reprinted from Ref. [67], with permission from the Royal Society of Chemistry.

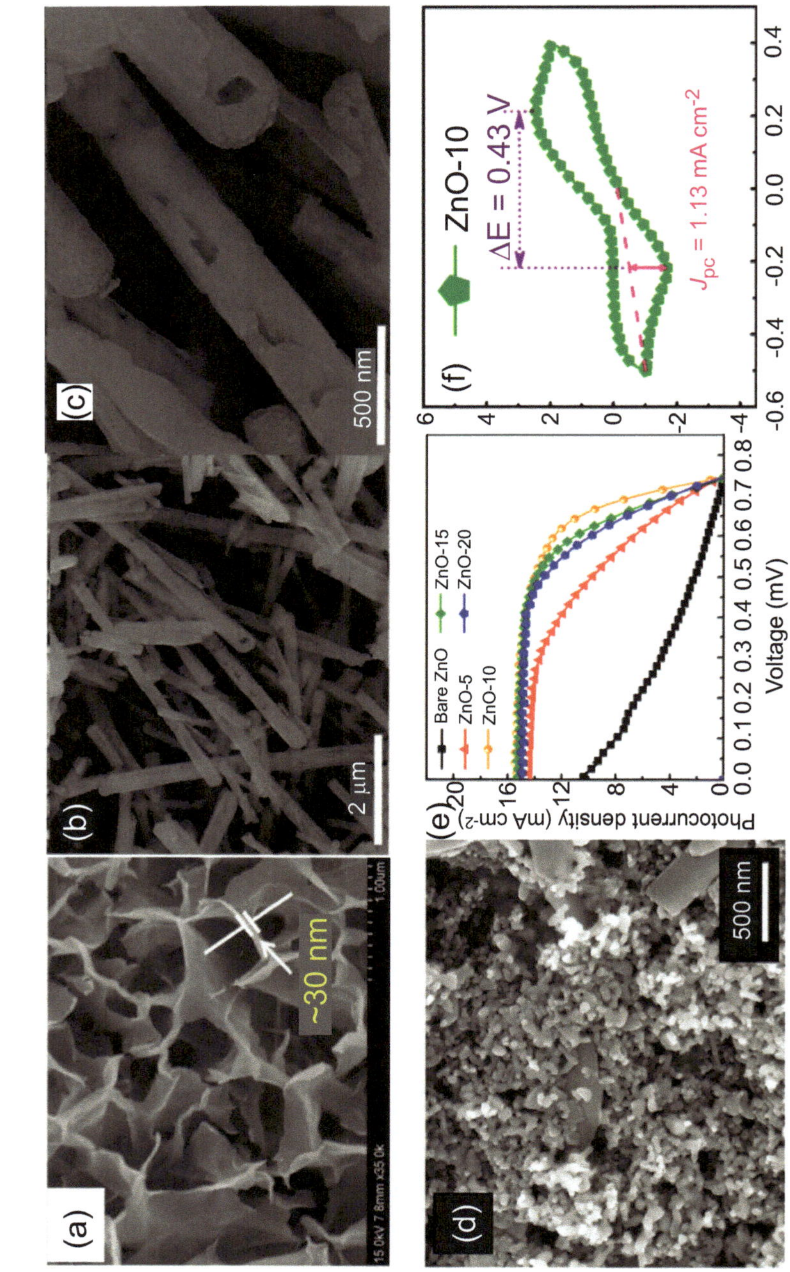
(a)
~30 nm
(b)
2 µm
(c)
500 nm
(d)
500 nm
(e)
Bare ZnO
ZnO-5
ZnO-10
ZnO-15
ZnO-20
Photocurrent density (mA cm-2)
Voltage (mV)
(f)
ZnO-10
ΔE = 0.43 V
Jpc = 1.13 mA cm-2

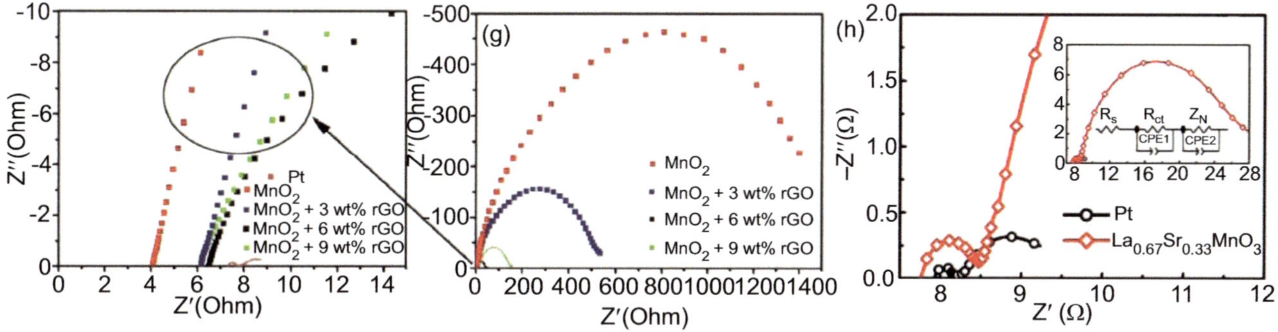

Figure 2.11 SEM images of (a) NiO thin film at 1 μm magnification, reprinted from Ref. [69], with permission from Springer Nature; and (b, c) MnO_2, reprinted from Ref. [70], with permission from the Royal Society of Chemistry. (d) FE-SEM images of three-dimensional porous structure of ZnO/PEDOT:PSS composites, reprinted from Ref. [71], with permission from American Chemical Society. (e) *J–V* curves of DSSCs with bare ZnO and ZnO/PEDOT:PSS composites CE, reprinted from Ref. [71], with permission from American Chemical Society. (f) Cyclic voltammograms of the films of ZnO/PEDOT:PSS composites CE, reprinted from Ref. [71], with permission from American Chemical Society. (g) Nyquist plots of the symmetrical cells fabricated with Pt and four MnO_2-based CEs, reprinted from Ref. [70], with permission from the Royal Society of Chemistry. (h) Nyquist plots for the symmetrical cells fabricated with LSMO ($La_{0.67}Sr_{0.33}MnO_3$) and Pt electrodes, reprinted from Ref. [75], with permission from AIP Publishing.

P. Jin et al. [70] prepared CE for DSSCs using MnO_2 nanotubes (Fig. 2.11b,c) with assistance of graphene. The electrochemical catalysis of MnO_2 was excellent, but the EIS value (Fig. 2.11g) of MnO_2 was too large. This EIS value was reduced by adding rGO because it has high Fermi velocity, high electrical conductivity, structural flexibility, and linear energy dispersion relation. The combination of MnO_2 nanotubes with rGO CE provides electrochemical stability and conductivity. Different weight percentages of rGO were added one by one and they reduced the EIS value. MnO_2 with 6 wt% rGO CE provides PCE (5.77%) higher than the conventional Pt CE PCE (5.40%).

2.7.7 3D Morphology-Based CE

C.-T. Li et al. [71] prepared (Fig. 2.11d) three-dimensional porous structured of ZnO/PEDOT:PSS composites. This composite material has high surface roughness and gives enough electrocatalytic active sites for the reduction of I_3^-. The CV curves (Fig. 2.11f) show the cathodic peak current density (J_P) and peak potential separation (ΔE). The ZnO CE shows the higher J_P values compared to polymer matrix, which indicates the electrocatalytic ability. The ZnO CE has a small ΔE value, but it can be rectified by incorporating into PEDOT:PSS polymer matrix. ZnO composite CE shows the PCE (7.54%) closer to the conventional Pt (8.59%) (Fig. 2.11e).

2.7.8 Other Morphology-Based CEs

S. Yun et al. [72] prepared TaO_x, which is used as high electrocatalytic CE for the DSSC application. This TaO_x CE exhibits the PCE value (6.79%) very close to Pt CE PCE (7.57%). G. R. Mutta et al. [73] prepared MoO_3 thin films by the sol–gel method for the alternative CE for DSSCs. The PCE of MoO_3 CE (1.04%) is lower than the Pt CE (7%). However, they introduced these materials as new materials for CE purpose. L. Wang et al. [74] prepared NbO_2 CE, which has small R_{ct}, large R_s, and a large surface area that give better electrocatalytic activity than Pt CE. The PCE of NbO_2 CE (3.62%) is better than the Pt CE PCE (2.91%). Y. Zhong et al. [75] prepared ternary perovskite type $La_{0.67}Sr_{0.33}MnO_3$ (LSMO) CE for DSSC. The electrochemical studies (Fig. 2.11h) indicate that LSMO has an excellent electrocatalytic activity for triiodide reduction and excellent electrochemical stability for the I^-/I_3^- electrode system. The LSMO CE PCE (6.62%) is very close to the Pt CE PCE (7.15%). M. Wu et al. [76] prepared SnO_2 and Nb_2O_5 CE by sintering at air and N_2 atmosphere. The N_2 atmosphere CEs increase the PCE higher compared to air atmosphere CEs due to the effect of N_2 atmosphere. The catalytic variation results due to changes in surface chemical and physical properties of SnO_2 and Nb_2O_5 CE. As a result, electronic structure changes govern different catalytic activities.

2.7.9 Effects of Oxygen Vacancy in CE

J. Pan et al. [77] prepared a nonstoichiometric $SnO_{2-\delta}$ nanocrystal-based CE for DSSC. Here, they checked the improvement of PCE with ($SnO_{2-\delta}$) and without (SnO_2) creating oxygen vacancies by heating argon and air atmosphere. The oxygen vacancies are created inside the material, not on the surface. The $SnO_{2-\delta}$ R_{ct} value at the counter electrode/electrolyte interface is smaller than the SnO_2 CE due to oxygen vacancies. This indicates that the catalytic activity of $SnO_{2-\delta}$ is higher than that of SnO_2 CE. Therefore, the PCE of $SnO_{2-\delta}$ CE (4.81%) is also higher than that of SnO_2 CE (2.88%). H. Elbohy et al. [78] prepared WO_{3-x} and SnO_{2-x} CE for DSSC for creating oxygen vacancies by urea treatment and annealing through N_2 atmosphere. The pristine MO (WO_3 and SnO_2) delivers low PCE because of its poor catalytic activity for reduction of triiodide. The N_2 annealed metal oxide CE (WO_{3-x} and SnO_{2-x} CE) contributes good PCE due to the creation of oxygen vacancies or shell states that assist the charge transfer at CE/electrolyte interface. The overall PCE of these MOs (WO_{3-x} and SnO_{2-x} CE) is still lower than the Pt CE. So they mixed the urea content with different wt% linearly in pristine MO (WO_3 and SnO_2) to create more oxygen vacancies and more electronic shallow states due to the presence of H and N in urea. However, the linear increase of urea content did not increase the PCE after certain threshold or wt% because more increase of oxygen vacancies enhances the back recombination of injected electrons from CE to electrolyte, which produces low PCE. Similarly, P. Uppachai et al. [79] also prepared efficient WO_{3-x} CE. It exhibits excellent electrical conductivity and catalytic activity due to oxygen vacancy creation through annealing in Ar atmosphere. The PCE of WO_{3-x} CE (5.25%)

is small compared to Pt CE (6.96%). Photovoltaic performance of DSSCs using metal oxides and their composite CEs is summarized in Table 2.1.

Table 2.1 Photovoltaic performance of DSSC using metal oxide CEs

CE materials	J_{sc} (mA/cm^2)	V_{oc} (mV)	FF (%)	PCE (%)	PCE (Pt) (%)	References
α-Fe_2O_3	15.92	784	56	6.96	7.32	[46]
α-MnO_2 nanorods	14.7	650	38	4.1	2.91	[61]
γ-Fe_2O_3	15.0	690	61	6.39	7.93	[67]
$Bi_5FeTi_3O_{15}$	3.80	364	15.9	0.22	12.21	[65]
CuO	14.06	670	28	2.73	6.36	[33]
Cr_2O_3	8.46	788	16	1.07	7.50	[80]
Fe_3O_4	15.41	701	63	6.80	7.93	[67]
$La_{0.67}Sr_{0.33}MnO_3$	15.2	772	57	6.62	7.15	[75]
MoO_2	7.84	765	40	2.40	7.50	[80]
MoO_3	8.17	700	16.59	1.04	7	[73]
MnO_2	11.92	677	41.4	3.04	5.40	[70]
NbO_2	13.90	810	70	7.88	7.65	[45]
Nb_2O_5	9.98	808	60	4.84	7.50	[80]
NiO	4.83	799	07	0.28	8.7	[39]
RuO_2	16.51	813	54	7.22	7.17	[81]
SnO_2	20.1	720	44.2	6.44	8.45	[78]
SnO_{2-x}	19.92	750	58	8.7	8.45	[78]
TaO_x	13.17	760	68	6.79	7.57	[82]
TaO	12.59	770	67	6.48	7.16	[58]
Ta_2O_5	13.01	750	42	4.08	7.16	[58]
TiO_2	3.10	742	33	0.76	7.50	[80]
V_2O_3	10.99	780	63	5.40	7.50	[80]
V_2O_5	8.34	680	27	1.6	6.95	[56]
WO_2	14.02	808	64	7.25	7.57	[59]

CE materials	J_{sc} (mA/cm^2)	V_{oc} (mV)	FF (%)	PCE (%)	PCE (Pt) (%)	References
$WO_{2.72}$	14.90	770	70	8.03	8.08	[55]
WO_3	18.91	720	44	6.04	8.45	[78]
WO_{3-x}	20.51	740	56	8.57	8.45	[78]
$W_{18}O_{49}$ NFs	17.14	700	66	7.94	8.01	[66]
ZrO_2	11.88	782	28	2.60	7.50	[80]
BFTO/graphene	21.56	740	59.9	9.56	12.21	[65]
Co_3O_4-MC-Co_2N	16.26	688	45	5.08	4.88	[41]
CuO + GNs	15.62	690	31	3.40	6.36	[33]
MnO_2 + rGO	15.97	705	50.9	5.77	5.40	[70]
NiO-NF/MWCNT	18.54	644	63.9	7.63	6.72	[44]
NiO/PEDOT:PSS	22	748	46	7.58	8.7	[39]
Pt/NiO	10.44	650	63.15	4.28	3.16	[35]
Pt/SnO_2	16.24	731	74	8.83	8.04	[38]
PEDOT-GO/TiO_2	3.967	650	45.22	1.166	0.727	[36]
SiO_2/PEDOT–PSS	13.5	720	58.2	5.66	6.38	[40]
VO_2(M1)	4.67	610	43.72	1.25	7.08	[47]
WO_3/WO_2	12.17	809	61	6.07	7.65	[62]
WO_2-MC	13.55	808	71	7.76	7.55	[60]
ZnO/PEDOT:PSS	15.35	740	66	7.54	8.59	[71]
ZnO nanorods decorated graphene/ZnO NP composites	0.47	830	43.6	0.17	0.15	[54]

2.8 Summary

This chapter discusses the introduction of DSSCs and their importance in the present situation. It explains the importance of metal oxides and their composites in DSSCs and their role in CEs. The chapter discusses the various methods of metal oxide preparation. After that, it describes metal oxide composite CEs and their role in enhancing PCE in DSSCs. Then it discusses the structural, morphological, and bandgap roles of metal oxides and their composites in CEs and their effects on PCE. It also explains the effects of oxygen vacancies in CEs. Various metal oxides and their composite CEs are found to give higher or equal PCE compared to the conventional Pt CE PCE.

Acknowledgment

The author Rajkumar C. expresses his gratitude to University Grants Commission (UGC) for providing UGC fellowship.

References

1. D. Sengupta, P. Das, B. Mondal, K. Mukherjee, Effects of doping, morphology and film-thickness of photo-anode materials for dye sensitized solar cell application: A review, *Renew. Sustain. Energy Rev.*, 60 (2016) 356–376. doi:10.1016/j.rser.2016.01.104.
2. J. Wu, Z. Lan, J. Lin, M. Huang, Y. Huang, L. Fan, G. Luo, Y. Lin, Y. Xie, Y. Wei, Counter electrodes in dye-sensitized solar cells, *Chem. Soc. Rev.*, 46 (2017) 6023. doi:10.1039/c6cs00752j.
3. A. Arulraj, M. Ramesh, B. Subramanian, G. Senguttuvan, In-situ temperature and thickness control grown 2D-MoS_2 via pulsed laser ablation for photovoltaic devices, *Sol. Energy*, 174 (2018) 286–295. doi:10.1016/j.solener.2018.08.056.
4. A. Arulraj, B. Subramanian, M. Ramesh, G. Senguttuvan, Effect of active sites in pulsed laser deposited bimetallic $NiMoS_2$ thin films for solar energy conversion, *Mater. Lett.*, 241 (2019) 132–135. doi:10.1016/j.matlet.2019.01.068.
5. R. Karthick, A. Arulraj, M. Ramesh, M. Selvaraj, Free-standing graphene/NiMoS paper as cathode for quasi-solid state dye-sensitized solar cells, *J. Colloid Interf. Sci.*, 530 (2018) 179–188. doi.org/10.1016/j.jcis.2018.06.082S.

6. A. Arulraj, S. Govindan, S. Vadivel, B. Subramanian, Photovoltaic performance of TiO_2 using natural sensitizer extracted from *Phyllanthus reticulatus*, *J. Mater. Sci. Mater. Electron.*, 28 (2017) 18455–18462. doi:10.1007/s10854-017-7792-7.
7. C. Rajkumar, A. Arulraj, Light scattering materials as photoanodes, in: *Interfacial Engineering in Functional Materials for Dye-Sensitized Solar Cells*, Wiley, 2019: pp. 79–105. doi:10.1002/9781119557401.ch4.
8. A. Arulraj, G. Senguttuvan, S. Veeramani, V. Sivakumar, B. Subramanian, Photovoltaic performance of natural metal free photo-sensitizer for TiO_2 based dye-sensitized solar cells, *Optik (Stuttg)*, 181 (2019) 619–626. doi:10.1016/j.ijleo.2018.12.104.
9. C. Rajkumar, A. Arulraj, Seed mediated synthesis of nanosized zinc oxide and its electron transporting activity in dye-sensitized solar cells, *Mater. Res. Express*, 5 (2018) 015029. doi:10.1088/2053-1591/aaa2b3.
10. U. Ahmed, M. Alizadeh, N. A. Rahim, S. Shahabuddin, M. S. Ahmed, A. K. Pandey, A comprehensive review on counter electrodes for dye sensitized solar cells: A special focus on Pt-TCO free counter electrodes, *Sol. Energy*, 174 (2018) 1097–1125. doi:10.1016/j.solener.2018.10.010.
11. M. Wu, T. Ma, Recent progress of counter electrode catalysts in dye-sensitized solar cells, *J. Phys. Chem. C*, 118 (2014) 16727–16742. doi:10.1021/jp412713h.
12. R. Irani, N. Naseri, S. Beke, A review of 2D-based counter electrodes applied in solar-assisted devices, *Coord. Chem. Rev.*, 324 (2016) 54–81. doi:10.1016/j.ccr.2016.06.010.
13. R. Jose, V. Thavasi, S. Ramakrishna, Metal oxides for dye-sensitized solar cells, *J. Am. Ceram. Soc.*, 92 (2009) 289–301. doi:10.1111/j.1551-2916.2008.02870.x.
14. M. V. Landau, Transition metal oxides, *Annu. Rev. Phys. Chem.*, 34 (2003) 2789. doi:10.1002/chin.200326249.
15. J. Theerthagiri, R. Senthil, J. Madhavan, A. R. Senthil, J. Madhavan, T. Maiyalagan, Recent progress in non-platinum counter electrode materials for dye-sensitized solar cells, *ChemElectroChem*, 2 (2015) 928–945. doi:10.1002/celc.201402406.
16. C. Gao, Q. Han, M. Wu, Review on transition metal compounds based counter electrode for dye-sensitized solar cells, *J. Energy Chem.*, 27 (2017) 703–712. doi:10.1016/j.jechem.2017.09.003.

17. A. Majid, M. Bibi, Wet chemical synthesis methods, in:*Cadmium Based II-VI Semiconducting Nanomaterials. Topics in Mining, Metallurgy and Materials Engineering.* Springer, Cham, 2018: pp. 43–101. doi:10.1007/978-3-319-68753-7_3.
18. C. Rajkumar, R. K. Srivastava, Time dependent rise and decay of photocurrent in zinc oxide nanoparticles in ambient and vacuum medium, *Mater. Res. Express*, 5 (2018) 055002. doi:10.1088/2053-1591/aabeca.
19. C. Rajkumar, Structural, optical and photoconductivity studies of ZnO bicones synthesized by seed-mediated method, *Vacuum*, 168 (2019) 108856. doi:10.1016/j.vacuum.2019.108856.
20. C. Rajkumar, R. K. Srivastava, UV–visible photoresponse properties of self-seeded and polymer mediated ZnO flower-like and biconical nanostructures, *Results Phys*, 15 (2019) 102647. doi:10.1016/j.rinp.2019.102647.
21. A. S. Razeen, A. Gadallah, Ultraviolet stimulated emission from sol-gel spin coated ZnO thin films, *Adv. Mater. Sci. Eng.*, 2017 (2017). doi:10.1155/2017/9464862.
22. W.-C. Lin, Y.-C. Lin, M. Esashi, A. A. Seshia, In-situ hydrothermal synthesis of zinc oxide nanostructures using microheaters, *IEEE Trans. Nanotechnol.*, 14 (2015) 1046–1053. doi:10.1109/TNANO.2015.2468076.
23. C. Hu, J. Mi, S. Shang, J. Shangguan, The study of thermal decomposition kinetics of zinc oxide formation from zinc oxalate dihydrate, *J. Therm. Anal. Calorim.*, 115 (2014) 1119–1125. doi:10.1007/s10973-013-3438-z.
24. K.-H. Habig, Chemical vapor deposition and physical vapor deposition coatings: Properties, tribological behavior, and applications, *J. Vac. Sci. Technol. A*, 4 (1986) 2832–2843. doi:10.1116/1.573687.
25. C. Satienpattanakoon, D. Yiamsawas, W. Kangwansupamonkon, R. Nuisin, Synthesis and characterization of zinc oxide nanocrystals by solid-state and solvothermal techniques, *Adv. Mater. Res.*, 55–57 (2008) 657–660. doi:10.4028/www.scientific.net/AMR.55-57.657.
26. S. Duo, Y. Li, H. Zhang, T. Liu, K. Wu, Z. Li, A facile salicylic acid assisted hydrothermal synthesis of different flower-like ZnO hierarchical architectures with optical and concentration-dependent photocatalytic properties, *Mater. Charact.*, 114 (2016) 185–196. doi:10.1016/j.matchar.2016.02.021.

27. B. Gersten, Solvothermal synthesis of nanoparticles, *Chemfiles*, 5 (2018) 13. http://www.sigmaaldrich.com/technical-documents/articles/chemfiles/solvothermal-synthesis.html.

28. J. N. Tiwari, R. N. Tiwari, K. S. Kim, Zero-dimensional, one-dimensional, two-dimensional and three-dimensional nanostructured materials for advanced electrochemical energy devices, *Prog. Mater. Sci.*, 57 (2012) 724–803. doi:10.1016/j.pmatsci.2011.08.003.

29. S. Sakka, Sol-gel process and applications, in: *Handbook of Advanced Ceramics*, Second Edition, Elsevier, 2013: pp. 883–910. doi:10.1016/B978-0-12-385469-8.00048-4.

30. L. Znaidi, Sol-gel-deposited ZnO thin films: A review, *Mater. Sci. Eng. B*, 174 (2010) 18–30. doi:10.1016/j.mseb.2010.07.001.

31. H. Hayashi, Y. Hakuta, Hydrothermal synthesis of metal oxide nanoparticles in supercritical water, *Materials (Basel)*, 3 (2010) 3794–3817. doi:10.3390/ma3073794.

32. A. Ali, H. Zafar, M. Zia, I. ul Haq, A. R. Phull, J. S. Ali, A. Hussain, Synthesis, characterization, applications, and challenges of iron oxide nanoparticles, *Nanotechnol. Sci. Appl.*, 9 (2016) 49–67. doi:10.2147/NSA.S99986.

33. C.-H. Tsai, P.-H. Fei, C.-M. Lin, S.-L. Shiu, CuO and CuO/graphene nanostructured thin films as counter electrodes for Pt-free dye-sensitized solar cells, *Coatings*, 8 (2018) 21. doi:10.3390/coatings8010021.

34. J. Xia, C. Yuan, S. Yanagida, Novel counter electrode V_2O_5/Al for solid dye-sensitized solar cells, *ACS Appl. Mater. Interfaces*, 2 (2010) 2136–2139. doi:10.1021/am100380w.

35. C.-S. Chou, C. Hsiung, C. Wang, R. Yang, M. Guo, Preparation of a counter electrode with P-type NiO and its applications in dye-sensitized solar cell, *Int. J. Photoenergy*, 2010 (2010) 1–9. doi:10.1155/2010/902385.

36. M. N. Mustafa, S. Shafie, Z. Zainal, Y. Sulaiman, A novel poly(3,4-ethylenedioxythiophene)-graphene oxide/titanium dioxide composites counter electrode for dye-sensitized solar cell, *J. Nanomater.*, 2017 (2017) 4045672.

37. W. Cui, The preparation and performance of WO3@C as a counter electrode catalyst for dye-sensitized solar cell, *Int. J. Electrochem. Sci.*, 12 (2017) 11487–11495. doi:10.20964/2017.12.72.

38. X. Chen, Y. Hou, S. Yang, X. H. Yang, H. G. Yang, A novel strategy to prepare a Pt–SnO_2 nanocomposite as a highly efficient counter electrode for

dye-sensitized solar cells, *J. Mater. Chem. A*, 2 (2014) 17253–17257. doi:10.1039/C4TA03774J.

39. H. Wang, W. Wei, Y. H. Hu, NiO as an efficient counter electrode catalyst for dye-sensitized solar cells, *Top. Catal.*, 57 (2014) 607–611. doi:10.1007/s11244-013-0218-8.
40. D. Song, M. Li, Y. Li, X. Zhao, B. Jiang, Y. Jiang, Highly transparent and efficient counter electrode using SiO_2/PEDOT–PSS composite for bifacial dye-sensitized solar cells, *ACS Appl. Mater. Interfaces*, 6 (2014) 7126–7132. doi:10.1021/am500082x.
41. M. Chen, L. L. Shao, Z. M. Gao, T. Z. Ren, Z. Y. Yuan, Cobalt oxide and nitride particles supported on mesoporous carbons as composite electrocatalysts for dye-sensitized solar cells, *J. Power Sources*, 286 (2015) 82–90. doi:10.1016/j.jpowsour.2015.03.153.
42. M. R. Al-bahrani, W. Ahmad, H. F. Mehnane, Y. Chen, Z. Cheng, Y. H. Gao, Enhanced electrocatalytic activity by RGO/MWCNTs/NiO counter electrode for dye-sensitized solar cells, *Nano-Micro Lett.*, 7 (2015) 298–306. doi:10.1007/s40820-015-0043-7.
43. S. S. Kim, K. W. Park, J. H. Yum, Y. E. Sung, Dye-sensitized solar cells with Pt-NiO and Pt-TiO_2 biphase counter electrodes, *J. Photochem. Photobiol. A Chem.*, 189 (2007) 301–306. doi:10.1016/j.jphotochem.2007.02.019.
44. M. R. Al-Bahrani, L. Liu, W. Ahmad, J. Tao, F. Tu, Z. Cheng, Y. Gao, NiO-NF/MWCNT nanocomposite catalyst as a counter electrode for high performance dye-sensitized solar cells, *Appl. Surf. Sci.*, 331 (2015) 333–338. doi:10.1016/j.apsusc.2015.01.015.
45. X. Lin, M. Wu, Y. Wang, A. Hagfeldt, T. Ma, Novel counter electrode catalysts of niobium oxides supersede Pt for dye-sensitized solar cells, *Chem. Commun.*, 47 (2011) 11489–11491. doi:10.1039/c1cc14973c.
46. Y. Hou, D. Wang, X. H. Yang, W. Q. Fang, B. Zhang, H. F. Wang, G. Z. Lu, P. Hu, H. J. Zhao, H. G. Yang, Rational screening low-cost counter electrodes for dye-sensitized solar cells, *Nat. Commun.*, 4 (2013) 1583. doi:10.1038/ncomms2547.
47. G. R. Mutta, S. R. Popuri, M. Vasundhara, M. Maciejczyk, A. V. Racu, R. Banica, N. Robertson, J. I. B. Wilson, N. S. Bennett, Facile hydrothermal synthesis of economically viable VO_2(M1) counter electrode for dye sensitized solar cells, *Mater. Res. Bull.*, 83 (2016) 135–140. doi:10.1016/j.materresbull.2016.05.027.
48. M. W. Khan, J. Yao, K. Zhang, X. Zuo, Q. Yang, H. Tang, K. M. Ur Rehman, G. Li, M. Wu, K. Zhu, H. Zhang, Facile fabrication of hollow mesosphere of crystalline SnO_2 nanoparticles and synthesis of SnO_2@SWCNTs@

reduced graphene oxide nanocomposite as efficient Pt-free counter electrode for dye-sensitized solar cells, *Opt. Mater. (Amst)*, 80 (2018) 271–278. doi:10.1016/j.optmat.2018.04.059.

49. P. Vijayakumar, M. S. Pandian, S. Mukhopadhyay, P. Ramasamy, Synthesis and characterizations of large surface tungsten oxide nanoparticles as a novel counter electrode for dye-sensitized solar cell, *J. Sol-Gel Sci. Technol.*, 75 (2015) 487–494. doi:10.1007/s10971-015-3719-z.

50. Y. Yang, K. Ri, Y. Rong, L. Liu, T. Liu, M. Hu, X. Li, H. Han, Full-printable transparent monolithic solid-state dye-sensitized solar cell with mesoscopic indium tin oxide counter electrode, *Phys. Chem. Chem. Phys.*, 16 (2014) 17743–17747. doi:10.1039/c4cp02354d.

51. Y.-F. Chiang, C.-H. Tsai, P. Chen, T.-F. Guo, Bifacial transparent solid-state dye-sensitized solar cell with sputtered indium-tin-oxide counter electrode, *Sol. Energy*, 86 (2012) 1967–1972. doi:10.1016/j.solener.2012.03.004.

52. W. Ahmad, L. Chu, M. R. Al-Bahrani, X. Ren, J. Su, Y. Gao, P-type NiO nanoparticles enhanced acetylene black as efficient counter electrode for dye-sensitized solar cells, *Mater. Res. Bull.*, 67 (2015) 185–190. doi:10.1016/j.materresbull.2015.03.021.

53. H. Wang, W. Wei, Y. H. Hu, Efficient ZnO-based counter electrodes for dye-sensitized solar cells, *J. Mater. Chem. A*, 1 (2013) 6622–6628. doi:10.1039/c3ta10892a.

54. B.-G. Zhai, L. Yang, Y. M. Huang, ZnO nanorods decorated graphene/ZnO nanoparticle composite as the counter electrode of dye-sensitised solar cells, *Mater. Res. Innov.*, 19 (2015) s15–s20. doi:10.1179/1432891715Z.0000000002045.

55. H. Zhou, Y. Shi, L. Wang, H. Zhang, C. Zhao, A. Hagfeldt, T. Ma, Notable catalytic activity of oxygen-vacancy-rich $WO_{2.72}$ nanorod bundles as counter electrodes for dye-sensitized solar cells, *Chem. Commun.*, 49 (2013) 7626. doi:10.1039/c3cc44518f.

56. G. R. Mutta, S. R. Popuri, M. Maciejczyk, N. Robertson, M. Vasundhara, J. I. B. Wilson, N. S. Bennett, V_2O_5 as an inexpensive counter electrode for dye sensitized solar cells, *Mater. Res. Express*, 3 (2016) 035501. doi:10.1088/2053-1591/3/3/035501.

57. T. Balkan, A. S. Sarac, Morphological effect of composite TiO_2 nanorod-TiO_2 nanoparticle/PEDOT:PSS electrodes on triiodide reduction, *EXPRESS Polym. Lett.*, 11 (2017) 106–116.

58. S. Yun, L. Wang, W. Guo, T. Ma, Non-Pt counter electrode catalysts using tantalum oxide for low-cost dye-sensitized solar cells, *Electrochem. Commun.*, 24 (2012) 69–73. doi:10.1016/j.elecom.2012.08.008.

59. M. Wu, X. Lin, A. Hagfeldt, T. Ma, A novel catalyst of WO_2 nanorod for the counter electrode of dye-sensitized solar cells, *Chem. Commun.*, 47 (2011) 4535. doi:10.1039/c1cc10638d.

60. M. Wu, X. Lin, L. Wang, W. Guo, Y. Wang, J. Xiao, A. Hagfeldt, T. Ma, In situ synthesized economical tungsten dioxide imbedded in mesoporous carbon for dye-sensitized solar cells as counter electrode catalyst, *J. Phys. Chem. C*, 115 (2011) 22598–22602. doi:10.1021/jp205886d.

61. K. Ahmad, A. Mohammad, S. M. Mobin, Hydrothermally grown α-MnO_2 nanorods as highly efficient low cost counter-electrode material for dye-sensitized solar cells and electrochemical sensing applications, *Electrochim. Acta*, 252 (2017) 549–557. doi:10.1016/j.electacta.2017.09.010.

62. M. Wu, L. Mu, Y. Wang, Y. N. Lin, H. Guo, T. Ma, One-step synthesis of nano-scaled tungsten oxides and carbides for dye-sensitized solar cells as counter electrode catalysts, *J. Mater. Chem. A*, 1 (2013) 7519–7524. doi:10.1039/c3ta10628d.

63. S. Sui, Y. Liao, Y. Xie, X. Wang, L. Li, Z. Luo, W. Zhou, G. Wang, K. Pan, A. Cabot, High catalytic activity of $W_{18}O_{49}$ nanowire-reduced graphite oxide composite counter electrode for dye-sensitized solar cells, *ChemistrySelect*, 2 (2017) 8927–8935. doi:10.1002/slct.201701436.

64. C.-H. Tsai, P.-H. Fei, C.-H. Chen, Investigation of coral-like Cu_2O nano/microstructures as counter electrodes for dye-sensitized solar cells, *Materials (Basel)*, 8 (2015) 5715–5729. doi:10.3390/ma8095274.

65. H. W. Zheng, X. Liang, Y. H. Yu, K. Wang, X. A. Zhang, B. Q. Men, C. L. Diao, C. X. Peng, G. T. Yue, $Bi_5FeTi_3O_{15}$ nanofibers/graphene nanocomposites as an effective counter electrode for dye-sensitized solar cells, *Nanoscale Res. Lett.*, 12 (2017) 18. doi:10.1186/s11671-016-1799-5.

66. H. Zhou, Y. Shi, Q. Dong, Y. Wang, C. Zhu, L. Wang, N. Wang, Y. Wei, S. Tao, T. Ma, Interlaced $W_{18}O_{49}$ nanofibers as a superior catalyst for the counter electrode of highly efficient dye-sensitized solar cells, *J. Mater. Chem. A*, 2 (2014) 4347–4354. doi:10.1039/C3TA14345G.

67. L. Wang, Y. Shi, H. Zhang, X. Bai, Y. Wang, T. Ma, Iron oxide nanostructures as highly efficient heterogeneous catalysts for mesoscopic photovoltaics, *J. Mater. Chem. A*, 2 (2014) 15279–15283. doi:10.1039/c4ta03727h.

68. M. Shahpari, A. Behjat, M. Khajaminian, N. Torabi, The influence of morphology of hematite (α-Fe_2O_3) counter electrodes on the efficiency of dye-sensitized solar cells, *Sol. Energy*, 119 (2015) 45–53. doi:10.1016/j.solener.2015.06.039.

69. A. T. Yadav, P. P. Magar, V. S. Kadam, C. V. Jagtap, C. S. Pawar, Chemically deposited nickel oxide as counter electrode for dye sensitized solar cell, *J. Mater. Sci. Mater. Electron.*, 27 (2016) 12297–12301. doi:10.1007/s10854-016-5020-5.

70. P. Jin, X. Zhang, M. Zhen, J. Wang, MnO_2 nanotubes with graphene-assistance as low-cost counter-electrode materials in dye-sensitized solar cells, *RSC Adv.*, 6 (2016) 10938–10942. doi:10.1039/c5ra26995d.

71. C.-T. Li, H. Chang, Y. Li, Y. Huang, Y. Tsai, R. Vittal, Y. Sheng, K. Ho, Electrocatalytic zinc composites as the efficient counter electrodes of dye-sensitized solar cells: Study on the electrochemical performances and density functional theory calculations, *ACS Appl. Mater. Interfaces*, 7 (2015) 28254–28263. doi:10.1021/acsami.5b07724.

72. S. Yun, M. Wu, Y. Wang, J. Shi, X. Lin, A. Hagfeldt, T. Ma, Pt-like behavior of high-performance counter electrodes prepared from binary tantalum compounds showing high electrocatalytic activity for dye-sensitized solar cells, *ChemSusChem*, 6 (2013) 411–416. doi:10.1002/cssc.201200845.

73. G. R. Mutta, S. R. Popuri, J. I. B. Wilson, N. S. Bennett, Sol-gel spin coated well adhered MoO_3 thin films as an alternative counter electrode for dye sensitized solar cells, *Solid State Sci.*, 61 (2016) 84–88. doi:10.1016/j.solidstatesciences.2016.08.016.

74. L. Wang, E. W.-G. Diau, M. Wu, H.-P. Lu, T. Ma, Highly efficient catalysts for Co(II/III) redox couples in dye-sensitized solar cells, *Chem. Commun.*, 48 (2012) 2600. doi:10.1039/c2cc17389a.

75. Y. Zhong, P. Chen, B. Yang, X. Zuo, L. Zhou, X. Yang, G. Li, Low-cost platinum-free counter electrode of $La_{0.67}Sr_{0.33}MnO_3$ perovskite for efficient dye-sensitized solar cells, *Appl. Phys. Lett.*, 106 (2015) 263903. doi:10.1063/1.4926339.

76. M. Wu, X. Lin, W. Guo, Y. Wang, L. Chu, T. Ma, K. Wu, Great improvement of catalytic activity of oxide counter electrodes fabricated in N_2 atmosphere for dye-sensitized solar cells, *Chem. Commun.*, 49 (2013) 1058. doi:10.1039/c2cc37360b.

77. J. Pan, L. Wang, J. C. Yu, G. Liu, H.-M. Cheng, A nonstoichiometric $SnO_{2-\delta}$ nanocrystal-based counter electrode for remarkably improving the performance of dye-sensitized solar cells, *Chem. Commun.*, 50 (2014) 7020. doi:10.1039/c4cc02066a.

78. H. Elbohy, K. M. Reza, E. Gamal, S. Abdulkarim, A. Gurung, G. Varnekar, M. H. Sayyad, Q. Qiao, Urea treated WO_3 and SnO_2 as cost effective and efficient counter electrodes of dye sensitized solar cells, in: *2016 IEEE*

International Conference on Electro Information Technology (EIT), 2016: pp. 477–481. doi:10.1109/EIT.2016.7535287.

79. P. Uppachai, V. Harnchana, S. Pimanpang, V. Amornkitbamrung, A. P. Brown, R. M. D. Brydson, A substoichiometric tungsten oxide catalyst provides a sustainable and efficient counter electrode for dye-sensitized solar cells, *Electrochim. Acta*, 145 (2014) 27–33. doi:10.1016/j.electacta.2014.08.096.

80. M. Wu, X. Lin, Y. Wang, L. Wang, W. Guo, D. Qi, X. Peng, A. Hagfeldt, M. Gratzel, T. Ma, M. Grätzel, T. Ma, Economical Pt-free catalysts for counter electrodes of dye-sensitized solar cells, *J. Am. Chem. Soc.*, 134 (2012) 3419–3428. doi:10.1021/ja209657v.

81. Y. Hou, Z. P. Chen, D. Wang, B. Zhang, S. Yang, H. F. Wang, Highly electrocatalytic activity of RuO_2 nanocrystals for triiodide reduction in dye-sensitized solar cells, *Small*, 10 (2013) 484–492. doi:10.1002/smll.201300653.

82. S. Yun, M. Wu, Y. Wang, J. Shi, X. Lin, A. Hagfeldt, Pt-like behavior of high-performance counter electrodes prepared from binary tantalum compounds showing high electrocatalytic activity for dye-sensitized solar cells, *ChemSusChem*, 6 (2013) 411–416. doi:10.1002/cssc.201200845.

Chapter 3

Dye-Sensitized Solar Cells Configuration with Transition Metal Carbides as Counter Electrode

J. Theerthagiri,[a,b] Sunitha Salla,[c] Seung Jun Lee,[b] Gilberto Maia,[d] J. Madhavan,[e] and Myong Yong Choi[b]

[a]*Centre of Excellence for Energy Research, Sathyabama Institute of Science and Technology (Deemed to be University), Chennai 600119, Tamil Nadu, India*
[b]*Department of Chemistry and Research Institute of Natural Sciences, Gyeongsang National University, Jinju 52828, South Korea*
[c]*Department of Chemistry, Sathyabama Institute of Science and Technology (Deemed to be University), Chennai 600119, Tamil Nadu, India*
[d]*Institute of Chemistry, Universidade Federal de Mato Grosso do Sul, Av. Senador Filinto Muller, 1555, Campo Grande, MS 79074-460, Brazil*
[e]*Solar Energy Lab, Department of Chemistry, Thiruvalluvar University, Vellore 632115, Tamil Nadu, India*
j.theerthagiri@gmail.com, mychoi@gnu.ac.kr

An increasing demand for energy and environmental promise has sustained serious research into solar energy conversion from renewable energy resources. Among various solar cells, dye-sensitized solar cells (DSSCs) with different kinds of new blend of materials are compelling device that have promising abilities in the conversion of solar energy into electricity with an extension to

Counter Electrode for Dye-Sensitized Solar Cells
Edited by Alagarsamy Pandikumar, Ramesh Mohan, and Kandasamy Jothivnekatachalam

ISBN 978-981-4877-38-1 (Hardcover), 978-1-003-11077-4 (eBook)
www.jennystanford.com

address consistently energy demands. A typical DSSC comprises four important components, namely, a photoanode, dye as sensitizer, redox electrolyte, and a counter electrode (CE). The CE is a vital component, in which triiodide is reduced to iodide by electron flowing through the external circuit. A compelling methodology of enhancing the execution of a CE is to upgrade the power conversion efficiency and to decrease the cost of the device. In this chapter, we have focused fundamentally on the transition metal carbides (TMCs) as CE for I^-/I_3^- redox electrolyte-based DSSCs, and apart from this TMC materials for iodine-free redox electrolytes are also discussed. The stability of TMC as CEs in DSSCs is also presented.

3.1 Introduction

Sunlight is thought to be one of the perfect sustainable energy resources and broadly utilized in the construction and human living as a result of its endless supply without region limit. The utilization of solar (sun) energy is reached incredible attractiveness and the devices based on solar energy which entering the market has been typically inclined against three primary parameters of power conversion efficiency, fabrication coat, and long-term stability. The solar cell is a competent sustainable energy conversion device that converts solar energy into electrical energy, and it can be the potential answer for the future energy issue that humankind faces. Among different solar cells, DSSCs show particular focal points over other photovoltaic cells because of low fabrication cost, simple fabrication protocol, less toxic, transparency, high power conversion efficiency, and large-scale production [1–10]. Generally, typical DSSCs contain four important parts, namely, photoanode (semiconductor oxide used as photoanode and mostly TiO_2 utilized), dye as sensitizer, I^-/I_3^- redox electrolyte, and CE. As of late, numerous investigations on the individual parts of DSSCs have been conveyed with a point of reducing the fabrication costs and accomplishing high power conversion efficiency. Among the various parts of DSSCs, the CE plays out a component of transferring photogenerated electrons from the outer circuit back to the redox shuttle and electrocatalytic reduction of I_3^- into I^- ions. In such manner, platinum (Pt) is favored because of its excellent electrocatalytic activity towards the reduction of I_3^- into I^- ions. However, it limits the large-scale applications because Pt is a very expensive rare earth metal and also undergoes slow dissolution in liquid electrolytes due to corrosive I_3^-/I^- redox couple, which deteriorates its long-life stability. These drawbacks have prompted numerous scientists to develop Pt-free CEs with low price and high electrocatalytic activity. To date, numerous functional materials have been utilized as Pt-free CE electrocatalysts, such as carbon materials, conducting polymers, metal sulfides, selenides, nitrides, and metal carbides [11–15].

Among these functional materials, TMCs have gained immense consideration as CEs due to their excellent electrocatalytic activity towards the reduction of I_3^- into I^- ions and unique physical and chemical properties when compared with transition metal nitrides and oxides with regard to various characteristics, such as high electrical conductivity, wear resistance, good thermal and chemical stability, high mechanical strength, and extreme hardness. Progress of metal carbides has indicated respectable Pt-like behavior owing to their similarity in the electronic and electrocatalytic properties with those of Pt-based materials. Nevertheless, a noteworthy challenge in this application is the production of metal carbides in single phase, high purity with homogeneously distributed nanosized particles. Based on precursor nature and carburization conditions, pure carbide materials with well structurally defined materials can be prepared. Commonly, metal carbides are prepared using temperature programmed reaction (TPR) process. In the TPR method, the precursors of corresponding metal oxides are thermally converted into carbides by high-temperature treatment with CH_4 and H_2. During TPR process, the precursor metal oxide is gradually reduced to lower valence oxide states and carburization is further undertaken to produce the required metal carbides. The synthetic challenges of metal carbide by utilizing CH_4 (g) process incorporates trouble in controlling the content of carbon. Consequently, there has been extensive enthusiasm for the advancement of new courses for the production of metal carbide materials that are most practical, steady, and less toxic with an extensive variety of applications [14, 16].

This chapter discusses the development of TMCs for the replacement of Pt CEs towards the electrocatalytic reduction of I_3^- into I^- ions in DSSCs. The advantages of metal carbides for the applications in DSSCs are also described.

3.2 Transition Metal Carbides as CEs for DSSCs

3.2.1 Tungsten Carbides

Tungsten carbide is a flexible material that exhibits good hardness, conductivity, a high melting point, durability, and electrocatalytic activity. Jang et al. [17] fabricated mesoporous WC as CE for DSSCs and showed the power conversion efficiency of 7.01% under 100 mW/cm^2, AM 1.5G illumination, which corresponded to 85% of the performance of DSSC with Pt CE. Wu et al. [18] prepared WC by the simple one-step chemical method and utilized it as CE for DSSCs. The fabricated DSSCs using WC/W_2C and W_2C exhibited the highest efficiency of 6.23% and 6.68%, respectively, comparable to that of the DSSC fabricated with the conventional Pt CE. Vijayakumar et al. [19] synthesized WC at various temperatures ranging from 800 °C to 1000 °C and employed as CE for DSSCs. The electrochemical investigation of WC revealed that increasing the sintering temperature leads to the increase in the charge transfer resistance and results in the decrease in the electrocatalytic activity of WC towards the reduction of the triiodide. The fabricated DSSC with WC nanorods sintered at 800 °C exhibited high power conversion efficiency of 0.41% compared to WC sintered at 900 °C and 1000 °C. Vijayakumar et al. [20] synthesized WC nanorods by pseudomorphic transformation of W_3O_8 nanorods using a high-temperature process. The as-prepared WC was employed as CE in DSSCs to replace the expensive Pt CE. The photovoltaic properties of the fabricated DSSCs with WC were examined under simulated standard global AM 1.5G sunlight (100 mW/cm^2). The overall power conversion efficiency of the fabricated DSSCs with binder and without binder using WC as CE was found to be 1.92% and 0.59%, respectively.

Jeong et al. [21] prepared WC/C composite nanofibers through electrospinning, followed by heat treatment using in situ carburization. The prepared composite was employed as the electrocatalyst for the reduction of I_3^- into I^- ions in I_3^-/I^- and disulfide/thiolate (T_2/T^-) redox electrolytes. The result of electrocatalytic activity of the WC/C was investigated over cyclic voltammetry (Fig. 3.1). The fabricated DSSCs with WC/C CE using I_3^-/I^- electrolytes exhibited the highest efficiency of 7.77%, which was 96% of that of the conventional Pt CEs. The DSSC fabricated with T_2/T^- electrolyte achieved the power conversion efficiency of

5.85%, which was higher than that of Pt CE (3.07%). The improved electrocatalytic performance of the WC/C nanofibers over the pristine materials was due to the positive synergetic effects, where the carbon acts as a charge transport pathway. Li et al. [22] developed the mesoporous carbon (MC) imbedded W_2C composite for the utilization of CEs in DSSCs. The W_2C/MC composite exhibited remarkable electrocatalytic performances towards the reduction of triiodide in organic T_2/T^- and conventional I_3^-/I^- redox couples, with a higher power conversion efficiency than that of the pristine W_2C and MC sample.

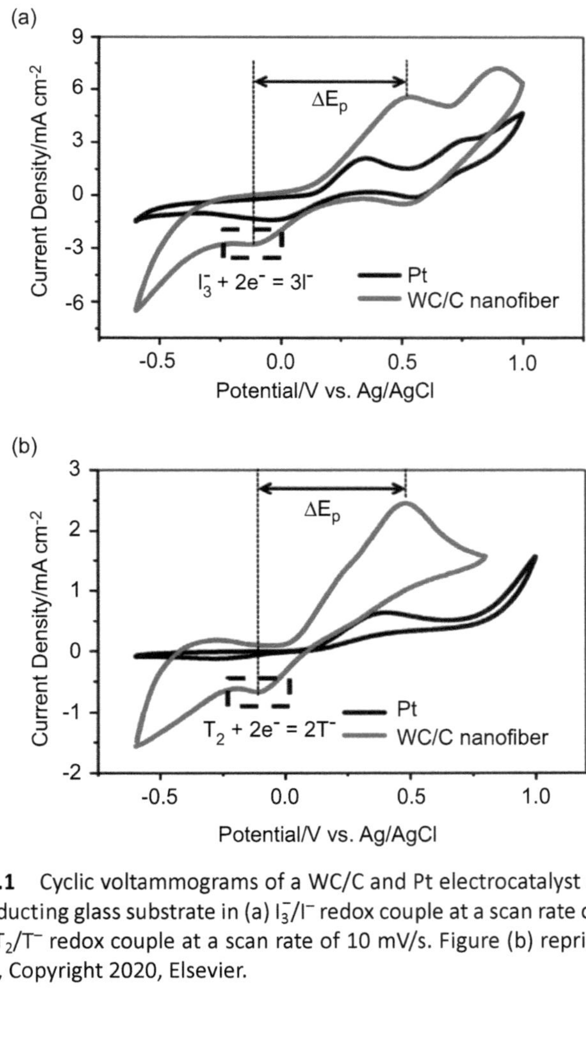

Figure 3.1 Cyclic voltammograms of a WC/C and Pt electrocatalyst coated on FTO conducting glass substrate in (a) I_3^-/I^- redox couple at a scan rate of 50 mV/s and (b) T_2/T^- redox couple at a scan rate of 10 mV/s. Figure (b) reprinted from Ref. [21], Copyright 2020, Elsevier.

3.2.2 Molybdenum Carbides

Wu et al. [23] prepared Mo_2C and utilized as CE for DSSCs. The fabricated DSSCs exhibited a better electrocatalytic performance for regeneration of an organic sulphide redox shuttle and showed the high power conversion efficiency of about 6.22%. Zheng et al. [24] prepared Mo_2C with two different structures of microsphere and nanorods for use as CE in DSSCs, especially in the regeneration of iodide-free redox couples of T_2/T^- and $Co^{3+}/^{2+}$. The surface morphologies of the prepared Mo_2C are shown in Fig. 3.2. In the T_2/ T^- based DSSCs, the prepared Mo_2C-Ms and Mo_2C-Nr both showed high catalytic property, and the devices generated power conversion efficiency values of 5.50% (Mo_2C-Ms) and 4.86% (Mo_2C-Ms), which were higher than the Pt CE (3.73%). In their investigation, they reported that Mo_2C-Ms shows higher catalytic activity for the $Co^{3+}/^{2+}$ redox couple that is competitive with the conventional Pt CE. The device using Mo_2C-Ms CE obtained a high PCE of 8.70%, indicating a great potential to replace the expensive Pt CE. The nanowire-shaped α-MoO_3 and various metal-doped (W, Nb, and Ta) α-MoO_3 were synthesized by Theerthagiri et al. [25] using the hydrothermal route. The synthesized metal oxides were converted

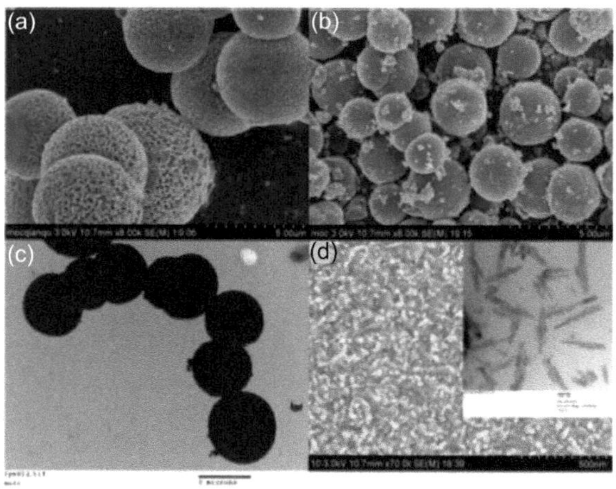

Figure 3.2 (a) Scanning electron microscope (SEM) image of molybdenum precursor, (b) SEM, (c) transmission electron microscope (TEM) images of Mo_2C microspheres, and (d) SEM image of Mo_2C nanorods; the insert shows the TEM image of Mo_2C nanorods. Figure (c) reproduced from Ref. [24], Copyright 2017, by permission of The Royal Society of Chemistry.

into corresponding carbides by the carburization method using urea as the carbon source. The order of electrocatalytic activity of the synthesized materials was W-α-Mo_2C > α-Mo_2C > Ta-α-Mo_2C > Nb-α-Mo_2C. The fabricated DSSCs with W-α-Mo_2C CE exhibited better power conversion efficiency (2.53%) than pure α-Mo_2C, Nb-, and Ta-doped α-Mo_2C. This study indicates that the W-doped α-Mo_2C is a potential electrocatalyst for applications in cost-effective DSSCs.

3.2.3 Titanium Carbides

Wu et al. [26] prepared titanium carbide/platinum (TiC/Pt) composite CEs using in situ chemical method for DSSCs. The DSSC using this composite CE yielded a high PCE of 7.63%, surpassing that of the Pt-based DSSC (7.16%). The composite TiC/Pt showed a good stability and after a long-term (1 year) durability test, the final power conversion efficiency value retained 86.9% of its highest efficiency. In the large-scale DSC module, TiC/Pt likewise showed a high catalytic activity and the DSC yielded a PCE of 4.98%, which was comparable to the performance of the DSC using the Pt CE (4.94%). Xie et al. [27] prepared the flexible TiC/carbon nanofiber (TiC/CNF) film by using electrospinning, followed by carbonization. These nanofibres had rough surface and random orientation, and the TiC nanoparticles (NPs) were uniformly dispersed on CNFs. The electrical conductivity increased, reaching the largest value, before decreasing as the amount of TiC NPs increased. These highly flexible films showed promising electrical conductivity and electrocatalytic activity and were used as CEs for flexible DSSCs. After the optimization of the amount of TiC NPs in CNFs (when the TiC content in precursor was 12 wt%), a high conversion efficiency of 3.80% was obtained. This was attributed to its optimum electrical conductivity and catalytic active sites. Wang et al. [28] prepared TiC film as CE for DSSCs and optimized the thickness of the film for effective performance. They reported in this investigation that the optimal film thickness of 20 μm of TiC CE exhibited high performance of 6.46%, which was comparable to the DSSCs fabricated with Pt CE. Yousef et al. [29] prepared cobalt–titanium carbide nanoparticles (Co-TiC NPs) embedded on carbon nanofibers by two-step process via electrospinning, followed by the carbonation process. The obtained composite was used as an efficient alternative to Pt-free CE for DSSCs. The fabricated electrode was shown to be effective, providing stable electrocatalytic activity towards the triiodide reduction. The J_{sc}, V_{oc}, FF, and η values were found to be ~9.98 mA/cm^2, 0.758 V, 0.507, and 3.87%, respectively.

3.2.4 Iron Carbides

Kim et al. [30] developed an active and economical material to replace Pt in DSSCs using nanostructured Fe_3C/carbon composite by electrochemical anodization of Fe foil, followed by heat treatment in carbon-bearing gas atmosphere. The fabricated DSSC using Fe_3C/carbon CE achieved an efficiency of 8.0%, which was slightly higher than that of the conventional Pt CE. Xu et al. [31] prepared hierarchical structured Fe_3C encaged in nitrogen-doped carbon (N-C) using a simple carbothermal reduction route and employed as CE for the DSSCs. The fabricated device exhibited an efficiency of 7.36%, which was comparable to that of the DSSC fabricated with Pt CE (7.15%). The good electrochemical performance was attributed to the synergistic effect of the combination of N-C and Fe_3C, which endows the nanocomposites with more interfacial active sites and improved electron transfer efficiency for the reduction of I_3^-/I^-. Liao et al. [32] prepared graphitic carbon (GC)/Fe_3C composite by a solid-state pyrolysis process and utilized as CE for DSSCs. The fabricated GC/Fe_3C composite exhibited better electrocatalytic performance than that of pure GC and Fe_3C. The enhanced electrocatalytic activity of the composite might be due to the positive synergistic effect and GC in the composite will help to rapid electron transport. The possible electrocatalytic mechanism of GC/Fe_3C electrocatalyst towards the reduction of triiodide is shown in Fig. 3.3. The DSSC based on GC/Fe_3C composite with 13.5 wt% Fe_3C content shows higher power conversion efficiency (6.04%), which was comparable to the DSSC using the conventional Pt CE (6.4%).

3.2.5 Vanadium Carbides

Vijayakumar et al. [33] developed VC CE and investigated using binder combination [conductive carbon/*N*-methylpyrrolidone (CC/NMP), NMP, CC/IPA (isopropanol)] material for increasing bonding strength, crack-free coated film functions, and so on on the FTO conducting glass substrate for the redox reaction at the CE of

DSSCs. The power conversion efficiency was 3.9% for VC–CC/NMP binder, which was comparable to the conventional Pt CE (4.0%). The enhanced performance was due to the CC/NMP binder, which increased the bonding strength between CE materials and FTO. Guo et al. [34] synthesized carbon supported TMC composites (Cr_3C_2-C, Mo_2C-C, WC-C, VC-C, NbC-C, TaC-C, and TiC-C). The carbon supported carbide composites show much higher electrocatalytic performances towards the cobalt redox couple regeneration in DSSCs as CE when compared to pure carbides. In the DSSC system, the devices using TiC-C, VC-C, and WC-C composite CEs display power conversion efficiencies of 8.85%, 9.75%, and 9.42%, respectively, which are much higher than those of the counterparts utilizing TiC, VC, and WC CEs.

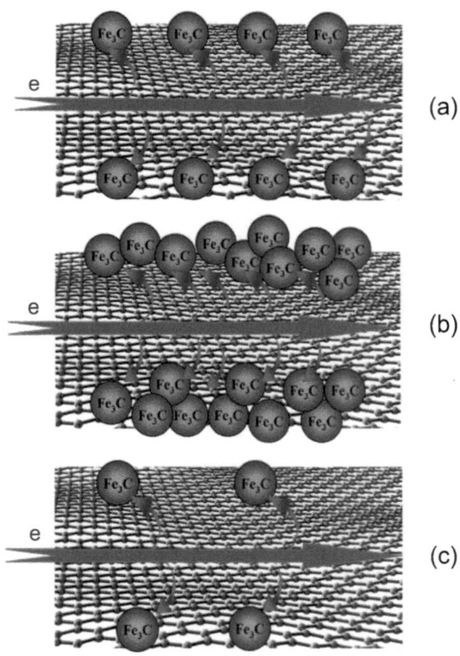

Figure 3.3 (a–c) Schematic view of possible electrocatalytic mechanism of GC/Fe_3C electrocatalyst towards the reduction of triiodide. Figure (c) reproduced from Ref. [32], Copyright 2018, by permission of The Royal Society of Chemistry.

The metal carbides are low-cost CE electrocatalysts for the reduction of I_3^- into I^- ions in DSSCs. The enhanced electrocatalytic activity of metal carbides might be due to the carbon atoms in metal carbide, which can alter the electronic structure of the parent metal to furnish a Pt-like d-band electronic structure.

3.3 Conclusions

In PV technologies, more focus has been given on DSSCs on account of their easy fabrication process, less toxic, less cost, and portable applications. Three principle parameters focused for DSSCs are their power conversion efficiency, stability, and cost-effectiveness. Pt is the generally used as the conventional CE electrocatalyst for DSSC applications because of its momentous electrocatalytic performance towards the triiodide reduction. In this chapter, we have focused fundamentally on TMCs as CE for I^-/I_3^- redox electrolyte-based DSSCs. A special focus has been placed on tungsten, molybdenum, iron, titanium, and vanadium carbides as alternative to Pt-free CE for DSSCs. The TMCs showed unique properties, such as high electrocatalytic activity towards the reduction of I_3^- into I^- ions, low cost, material diversity, prepared by simple synthesis, and ease of modification. Apart from this, TMC materials for iodine-free redox electrolytes are also discussed.

The commercial large-scale production of DSSC will required a large amount of Pt, which is very expensive and also less abundant in earth. Therefore, TMCs can be viewed as an efficient alternative to Pt CE for use in DSSCs. Further, the electrocatalytic performance of TMCs can be enhanced upon the advancement of synthesis methodologies resulting in better surface morphologies, smaller particle size, and higher surface area.

Acknowledgments

This work was supported by the Korea Basic Science Institute (KBSI) National Research Facilities & Equipment Center (NFEC) grant funded by the Korea government (Ministry of Education) (No. 2019R1A6C1010042). The authors Prof. M. Y. Choi and Dr. J. Theerthagiri acknowledge the financial support from National Research Foundation of Korea (NRF) (NRF-2019H1D3A1A01071209, NRF-2017M2B2A9A02049940).

References

1. S. Prasad, G. Durai, D. Devaraj, M.S. AlSalhi, J. Theerthagiri, P. Arunachalam, M. Gurulakshmi, M. Raghavender, P. Kuppusami, *RSC Adv.*, 8 (2018) 8828–8835.
2. J. Theerthagiri, R.A. Senthil, B. Senthilkumar, A.R. Polu, J. Madhavan, M. Ashokkumar, *J. Solid State Chem.*, 252 (2017) 43–71.
3. J. Theerthagiri, R.A. Senthil, M.H. Buraidah, K. Amarsingh Bhabu J. Madhavan, A.K. Arof, *Ionics*, 23 (2017) 1017–1025.
4. R.A. Senthil, J. Theerthagiri, J. Madhavan, S. Ganesan, A.K. Arof, *J. Phys. Chem. Solids*, 101 (2017) 18–24.
5. J. Theerthagiri, R.A. Senthil, P. Arunachalam, M.H. Buraidah, J. Madhavan, S. Amutha, A.K. Arof, *J. Solid State Electrochem.*, 21 (2017) 581–590.
6. J. Theerthagiri, R.A. Senthil, M.H. Buraidah, J. Madhavan, A.K. Arof, M. Ashokkumar, *J. Mater. Chem. A*, 4 (2016) 16119–16127.
7. J. Theerthagiri, R.A. Senthil, M.H. Buraidah, J. Madhavan, A.K. Arof, *J. Solid State Chem.*, 238 (2016) 113–120.
8. J. Theerthagiri, R.A. Senthil, M.H. Buraidah, A. Selvi, J. Madhavan, A.K. Arof, *Mater. Focus*, 5 (2016) 355–361.
9. J. Theerthagiri, R.A. Senthil, M.H. Buraidah, J. Madhavan, A.K. Arof, *Ionics*, 21 (2015) 2889–2896.
10. J. Theerthagiri, R.A. Senthil, M.H. Buraidah, J. Madhavan, A.K. Arof, *J. Appl. Polym. Sci.*, 132 (2015) 42489.
11. J. Theerthagiri, R.A. Senthil, J. Madhavan, T. Maiyalagan, *ChemElectroChem*, 2 (2015) 928–945.
12. J. Theerthagiri, R.A. Senthil, K. Susmitha, M. Raghavender, J. Madhavan, *Mater. Sci. Forum*, 832 (2015) 61–71.
13. R.A. Senthil, J. Theerthagiri, J. Madhavan, *J. Non-Cryst. Solids*, 406 (2014) 133–138.
14. J. Theerthagiri, R.A. Senthil, M.H. Buraidah, J. Madhavan, A.K. Arof, *J. Mater. Sci. Technol.*, 32 (2016) 1339–1344.
15. A. Hagfeldt, G. Boschloo, L. Sun, L. Kloo, H. Pettersson, *Chem. Rev.*, 110 (2010) 6595–6663.
16. M. Islam, R.M. Duarte, *Ceram. Int.*, 43 (2017) 10546–10553.
17. J.S. Jang, D.J. Ham, E. Ramasamy, J. Lee, J.S. Lee, *Chem. Commun.*, 46 (2010) 8600–8602.

18. M. Wu, L. Mu, Y. Wang, Y. Lin, H. Guo, T. Ma, *J. Mater. Chem. A*, 1 (2013) 7519–7524.
19. P. Vijayakumar, M. Senthil Pandian, S. P. Lim, A. Pandikumar, N. M. Huang, S. Mukhopadhyay, P. Ramasamy, *J. Mater. Sci. Mater. Electron.*, 26 (2015) 7977–7986.
20. P. Vijayakumar, M. Senthil Pandian, S.P. Lim, A. Pandikumar, N.M. Huang, S. Mukhopadhyay, P. Ramasamy, *Mater Sci. Semicond. Process.*, 39 (2015) 292–299.
21. I. Jeong, J. Lee, K.L.V. Joseph, H.I. Lee, J.K. Kim, S. Yoon, J. Lee, *Nano Energy*, 9 (2014) 392–400.
22. L. Li, H. Sui, W. Zhang, X. Li, K. Yang, A. Hagfeld, M. Wu, *J. Mater. Chem. C*, 4 (2016) 6778–6783.
23. M.X. Wu, Y.N. Lin, H. Guo, K. Wu, X. Lin, *Chem. Commun.*, 50 (2014) 7625.
24. H. Zheng, H. Guo, D. An, K. Wu, C. Gao, Q. Han, Y. Zhu, Y. Lin, M. Wu, *J. Mater. Chem. C*, 4 (2016) 6533–6538.
25. J. Theerthagiri, R.A. Senthil, M.H. Buraidah, J. Madhavan, A.K. Arof, *Mater. Today Proc.*, 3S (2016) S65–S72.
26. M.X. Wu, Y. Wang, X. Lin, W. Guo, K. Wu, Y.N. Lin, H. Guo, T. Ma, *J. Mater. Chem. A*, 1 (2013) 9672–9679.
27. X. Xie, X. Yin, J. Zhai, L. Song, P. Du, N. Li, J. Xiong, *J. Solid State Electrochem.*, 22 (2018) 1185–1195.
28. Y. Wang, M. Wu, X. Lin, A. Hagfeldt, T. Ma, *Eur. J. Inorg. Chem.*, 2012 (2012) 3557–3561.
29. A. Yousef, R.M. Brooks, M.H. El-Newehy, S.S. Al-Deyab, H.Y. Kim, *Int. J. Hydrogen Energy*, 42 (2017) 10407–10415.
30. J. Kim, J.S. Kang, J. Jeong, Y.J. Son, M.J. Lee, J. Kang, A. Lim, H.S. Park, Y.E. Sung, *J. Power Sources*, 396 (2018) 213–219.
31. H. Xu, C. Zhang, Z. Wang, S. Pang, X. Zhou, Z. Zhang, G. Cui, *J. Mater. Chem. A*, 2 (2014) 4676.
32. Y. Liao, K. Pan, L. Wang, Q. Pan, W. Zhou, X. Miao, B. Jiang, C. Tian, G. Tian, G. Wang, H. Fu, *ACS Appl. Mater. Interfaces*, 5 (2013) 3663–3670.
33. P. Vijayakumar, R. Govindaraj, N. Santhosh, M. Senthil Pandian, A. Pandikumar, P. Ramasamy, *J. Mater. Sci.*, 53 (2018) 4444–4455.
34. H. Guo, Q. Han, C. Gao, H. Zheng, Y. Zhu, M. Wu, *J. Power Sources*, 332 (2016) 399–405.

Chapter 4

Recent Advances in Transition Metal Nitrides Counter Electrode Based Dye-Sensitized Solar Cells

V. Gayathri,[a] I. John Peter,[a] P. Nithiananthi,[a] Smagul Karazhanov,[b] and C. Raja Mohan[a]

[a]*Nanostructure Lab., Department of Physics, The Gandhigram Rural Institute (Deemed to be University), Gandhigram 624302, Tamil Nadu, India*
[b]*Department for Solar Energy, Institute for Energy Technology, Kjeller 2027, Norway*
crmohan.gru@gmail.com, smagul.karazhanov@ife.no

An increasing demand for electricity leads researchers to utilize the solar energy through photovoltaics which is one among renewable energy resources and environmental friendly also. Among the various kinds of solar cells, the dye-sensitized solar cells (DSSCs) have received extensive attention due to their low fabrication cost, transparency, large-scale production, and moderately high efficiency. The platinum (Pt) counter electrode (CE) is used as crucial component in DSSC. Pt is a superior catalyst for reduction of I_3^- into I^- ions because of its high exchange current density, good catalytic activity, and transparency. Nowadays, the researcher views have focused on the replacement of expensive Pt with readily

Counter Electrode for Dye-Sensitized Solar Cells
Edited by Alagarsamy Pandikumar, Ramesh Mohan, and Kandasamy Jothivnekatachalam

ISBN 978-981-4877-38-1 (Hardcover), 978-1-003-11077-4 (eBook)
www.jennystanford.com

available and highly efficient electrocatalyst material. This chapter mainly discusses about the preparation of the metal nitrides and photovoltaic performance of metal nitrides versus Pt. Replacing the Pt CE by metal nitrides like MoN, TiN, CN, Zn_3N_2, VN, etc. have displayed solar cell efficiencies of about 7–9%.

4.1 Introduction

In solar cells power conversion efficiency (PCE) is primarily depending on the materials used in each component as photoanode, sensitizer, redox electrolyte, and counter electrode (CE). As a type of third generation solar cells, quantum dot-sensitized solar cells (QDSCs) are regarded as the derivation of dye-sensitized solar cells (DSSCs) and have attracted considerable attention of scientists worldwide. Since the outstanding work by O'Regan and Gratzel in 1991, the PCE of 14.3% [1, 2] has been achieved. However, high cost and complicated synthesis procedures of organic dyes with limited PCE, and durability restrict commercialization of DSSCs. Unlike DSSCs, QDSCs have advantages such as tunable bandgap, high extinction coefficients, large intrinsic dipole moment, and multiple exciton generation, which according to theoretical estimations [3–5] has potential to increase the conversion efficiency to as high as 44%. Commonly, QDs are made of II-VI, III-V, or IV-VI semiconductors such as CdS, CdSe, CdTe, PbS, InP, InAs, Ag_2S, Bi_2S_3, and ternary $CuInS_2$ [6–14]. Metal oxide semiconductors such as TiO_2 and ZnO are the commonly used electrodes in DSSC [15–18]. The electrolytes in DSSCs are organic electrolytes with iodide/triiodide (I^-/I_3^-) redox couple. In the past five years, the highest PCE ~5% to ~11.6% [19, 20] is reported. However, the I^-/I_3^- redox couple are corrosive to most QDs leading to the deterioration of QDSCs performance. Thus, it is critical to explore an appropriate non-corrosive electrolyte for stable QDSCs. In order to improve the performance of QDSCs, several electrolytes have been used that have been selected based on different redox couple such as the polysulfide (S^{2-}/Sx^{2-}), Fe^{2+}/Fe^{3+}, $Fe(CN)_6^{3-}/Fe(CN)_6^{4-}$, Co^{2+}/Co^{3+}, and $(CH_3)_4N_2S/(CH_3)_4N_2S_n$ [21–25]. Among them, QDSCs employed with polysulfide electrolyte have shown the highest short circuit current (J_{SC}) and PCE [26].

Counter electrode (CE) is an essential part of QDSCs that transfers electrons from the external circuit into the electrolyte and catalyzes the reduction reaction of the oxidized electrolytes at the electrolyte/CE interface. The impact of the CE on performance of QDSCs is derived from some admirable properties such as: (i) material conductivity, (ii) catalytic activity, as well as (iii) good chemical and mechanical stability [29]. Therefore, an appropriate CE material needs to display excellent electrocatalytic activity and good stability at the CE/electrolyte interface. Pt-based counter electrode (CE) has been widely studied in DSSCs device due to its excellent catalytic ability to reduce I_3^- to I^-. However, the Pt-based CEs are not suitable in a polysulfide-based QDSC for their strong chemisorption with polysulfide redox couple (S^{2-}/Sx^{2-}), leading to a reduced catalytic activity of CEs [27, 28]. A proper CE should catalyze regeneration of the redox in electrolyte and should take back the oxidized sensitizers to the ground state after catalysis thus keeping the device active and stable. Recently, significant efforts have been made to develop an alternative CE which includes noble metals (e.g., Au), conducting polymers [e.g., polyaniline, polypyrrole, and poly(3, 4-ethylenedioxythiophene)], metal chalcogenides (e.g., CuS, NiS, PbS, CoS, CuSe, FeS_2, Cu_2SnS_3, and $Cu_2ZnSnSe_4$), carbonaceous materials (e.g., carbon nanotubes, graphene, mesoporous carbon, activated carbon, carbon black, and carbon foams), as well as composite materials (e.g., CuS/CoS, CoS-NiS, and $NiCo_2S_4$).

Transition-metal carbides (TMC), nitrides (TMN), and oxides (TMO) have several advantages such as durability, high melting point, and electrical and thermal properties [30–33]. Thus, search of alternative catalytic materials with high conductivity to substitute conventional Pt-based CE is an urgent need to improve conversion efficiency. Recently TMN incorporated solar cells are found to deliver a higher performance due to its highest conductivity and electrocatalytic activity [34, 35]. This chapter focuses on latest developments and changes related to applications of transition metal nitrides such as **vanadium nitride (VN), titanium nitride (TN), molybdenum nitride (Mo_2N), tantalum nitride (Ta_3N_5), zinc nitride (Zn_3N_2),** etc. in dye-sensitized solar cells as counter electrode due to low cost and easy to prepare when compared to Pt. Moreover, the effect of size, shape on the electrocatalytic activity, and PCE are discussed.

Table 4.1 Physical and chemical properties of different nitrides and platinum

	Compound	VN	TiN	Mo_2N	Ta_3N_5	Zn_3N_2	Pt
1	**Physical state**	Solid	Solid	Solid	Solid	Solid	Solid
2	**Space group**	Fm3m	$Fm\bar{3}m$	$P\bar{3}ml$	$P\bar{6}2m$	la-3	Fm-3m
3	**Molar mass (g/mol)**	64.95	61.91	109.94	194.95	224.15	195.08
4	**Density (gm/cm^3)**	6.13	5.21	9.2	14.3	6.22	21.45
5	**Melting point**	2,320 °C	2,950 °C	645.87 °C	3,090 °C	Decomposes 700 °C	1770 °C
6	**Water solubility**	Insoluble	Insoluble	Insoluble	Insoluble	Insoluble	Insoluble
7	**Thermal conductivity**	-	29 W/(m·K)	-	-	-	71.6 W/(m·K)
8	**Toxicity**	Yes	No	Yes	No	No	No
9	**Reaction with**	Oxidizing agents	Oxidizing agents	Oxidizing agents	Oxidizing agents	Oxidizing agents	Oxidizing agents
10	**Flammability**	Non-flammable	Flammable	Non-flammable	Non-Flammable	Non-flammable	Non-flammable
12	**Oxidation number**	+3	+3	+3	+3	+2	+2 and +4
13	**Hydrogen bond donor count**	0	0	0	0	0	0
14	**Hydrogen bond acceptor count**	3	1	1	1	2	0
15	**Rotatable bond count**	0	0	0	0	0	0

Since 2010, several TMNs have been employed as CEs in DSSCs and here we discuss the synthesis techniques, effect of size, shape, and composition of TMNs on electrocatalytic activity and DSSC performance.

4.2 Vanadium Nitride (VN)

Mingxing Wu et al. synthesized three different VN nanostructures—VN peas, small and large VN cubes (S-VN, L-VN cubes)—and employed them into DSSC as CE catalysts [36]. Figure 4.1(a,b) shows the current density–voltage (*J–V*) spectra for S-VN, L-VN cubes, and VN peas using both (I_3^-/I^-) and disulfide/thiolate (T_2/T^-) medium. Among those nanostructures, VN peas delivered an open circuit voltage (V_{OC}) of 733 mV, short-circuit current (J_{SC}) of 15.06 mA/cm^2, fill factor (FF) of 0.66 and PCE of 7.29% in I_3^-/I^- electrolyte which is much higher than the L-VN cubes and S-VN cubes. The enhancement is owing to the performance of electrocatalytic activity. Compared to Pt-based CE, the T_2/T^-, the VN-peas-based CE shows the maximum PCE of 5.57%, an enhancement of 40.7%. The improvement is because of the larger active surface sites owing to the smaller particle size. Therefore, VN cubes have the larger activation energy for redox couple regeneration than VN peas. Figure 4.1(c,d) displays the CV curves of the VN-based materials, which reveals that the electrocatalytic performance is more enhanced for VN peas than that of other two electrodes.

Cathodic and anodic peak of the VN peas shifts in the positive and negative direction compared to those of VN cubes. It depicts the peak separation (ΔE_P). The obtained results show that, the VN peas acquired lower ΔE_P, which specifies better reversibility and it changes inversely with the charge transfer rate (k_s) hence achieved higher performance. The Nyquist plot (Fig. 4.1e) gives the electrochemical behavior at the electrode/electrolyte. The intercept on the real axis is due to the series resistance (R_S), the left and right arch gives the charge transfer resistance (R_{CT}) and the Nernst diffusion impedance (Z_N) of the electrode/electrolyte interface. No significant change was observed in the values of R_S and Z_N between the VN and Pt electrodes. But the R_{CT} value of VN peas and S-VN cubes are several times lower than that of Pt, indicating large reaction velocity. Undoubtedly, one can conclude that catalytic activity of VN depends on their size and shape.

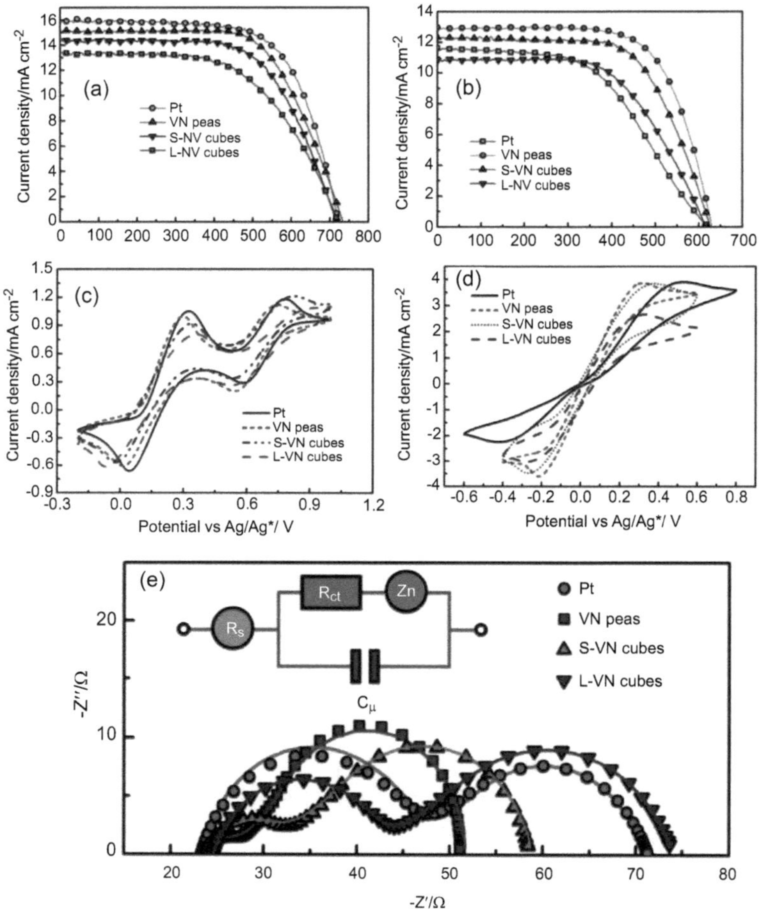

Figure 4.1 (a, b) *J–V* curves, (c, d) Cyclic voltammograms of the I^-/I_3^- and T_2/T^- electrolyte-based DSSC on different CEs, and (e) Nyquist plot. Reprinted with the permission from Ref. [36], Copyright 2014, American Chemical Society.

3D porous VN nanoribbons/reduced graphene oxide (PVNN/RGO) composite prepared by Guiqiang Wang et al. [37] and studied performance in DSSC which is displayed in Fig. 4.2. PVNN/RGO device achieved V_{OC} of 0.71 V, J_{SC} of 16.52 mA/cm^2, FF of 0.66 and PCE of 7.74%. The obtained PCE is larger than the pristine RGO and VN. The observed enhancement is because of high conductivity, lower R_{CT} and higher surface-active area, which enhanced the electrocatalytic performance. Performance of DSSC with VN-nanofibers (VN-NF) synthesized by Vijayakumar et al., [38] gives PCE of 6.8 %. Xiaoying Zhang et al. [39] fabricated DSSC using pristine nitride which gives low performances (2.463% for VN, 3.078% for TiN, and 6.179% for MoN), because of their low FF 28.75% for VN, 30.72% for TiN, and 58.98% for MoN. The poor adhesion of NG sheets was also one important reason for the N-doped reduced graphene oxide (NG) shows a low fill factor. The photovoltaic performance was increased by combining these two kinds of materials. The FF was amplified by 20% for MoN–NG, 120% for TiN–NG and 122% for VN–NG because of combining of nitrogen-doped reduced graphene oxide with the corresponding nitride. Therefore, incorporation of NG with nitrides shows a better power conversion efficiency than either constituent could alone. The results of some of VN-based DSSC are compared in Table 4.2.

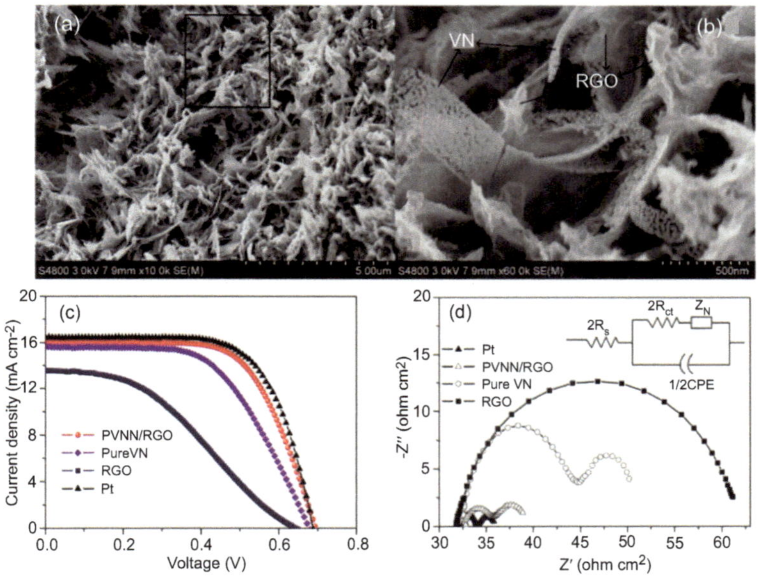

Figure 4.2 (a, b) SEM images of the as-prepared 3D PVNN/RGO composite. (c, d) JV and Nyquist plots of the cells fabricated with pure VN, RGO, PVNN/RGO, and Pt electrodes. Reprinted with the permission from Ref. [37], Copyright 2018, The Royal Society of Chemistry.

Table 4.2 DSSC performance for using VN as a CE

Counter electrode	Sensitizer	Photo-anode	J_{SC} (mA/cm^2)	V_{OC} (V)	FF	PCE (%)	Ref.
VN-NFs	N719	TiO_2	14.78	0.71	0.65	6.8	[38]
Pt			19.64	0.71	0.61	8.5	
VN-NG			12.58	0.78	0.63	6.2	[39]
Pt			14.22	0.79	0.69	7.8	
VN			11.74	0.78	0.64	5.6	[40]
Pt			14.08	0.78	0.68	7.5	

4.3 Molybdenum Nitride (MoN)

The performance of a counter electrode is greatly altered by apart from the VN, to discover an alternative counter electrode with good catalytic activity. Also, large mass transport is one of the major problems needing urgent attention. It has been reported that MoN have a high intrinsic electrocatalytic property, so it can behave equally and also better than Pt. X. P. Gao et al. [41] observed and cured these issues through optimizing the structure of MoN. They observed the smaller diffusion resistance with higher diffusion kinetics for MoN nanorod (NR)-Ti compared to MoN nanoparticle (NP)-Ti. The DSSC performance results are displayed in Fig. 4.3a. The DSSC using MoN NR-Ti CE possesses nearly similar performance to that using a Pt-FTO with V_{OC} of 0.740 V, J_{SC} of 15.26 mA/cm^2, and PCE of 7.29%.

From these results it is concluded [41] that MoN NR-Ti shows better performance than MoN-based NP-Ti. Moreover, the value for Z_N in the MoN NR electrode is 16.6 Ω whereas the result in the MoN NP electrode is 49.1 Ω. The lower ohmic polarization, the faster triiodide reduction, and higher exchange current density were the main reasons for increasing *V*oc and *J*sc. This work provides an effective scheme to manage the ion diffusion problem by manipulating and optimizing electrode structure. In contrast to optimizing the structure, X. P. Gao et al. and T. Ma et al. have prepared MoN, WN, and Fe_2N and studied their influence on DSSCs. Among them, MoN has superior electrocatalytic activity and a higher PCE [42, 43]. In addition, Song et al. employed MoN–carbon nanotubes

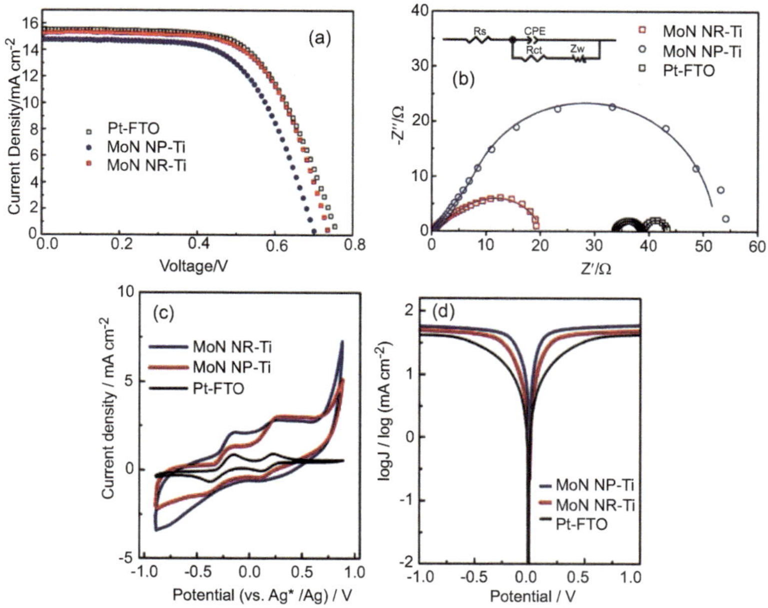

Figure 4.3 (a, b) Characteristic *J–V* and Nyquist curves. (c, d) Cyclic voltammogram and Tafel polarization curves with MoN NR-Ti, MoN NP-Ti, and Pt. Reprinted with the permission from Ref. [41], Copyright 2014, The Royal Society of Chemistry.

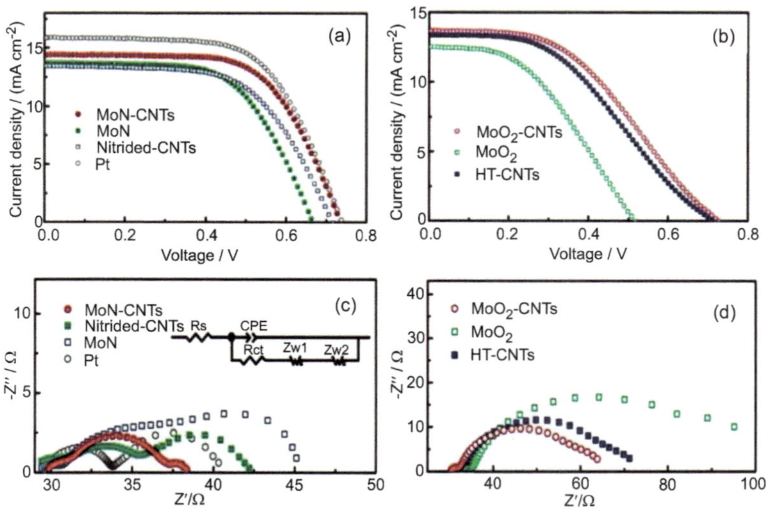

Figure 4.4 (a, b) Characteristic *J–V* and (c, d) Nyquist curves of DSSC using different CE. Reprinted with the permission from Ref. [44], Copyright 2012, The Royal Society of Chemistry.

(CNTs) composite as a CE for DSSC and got better performances as compared with the individual MoN [44]. In particular, the V_{OC} obtained for 0.735 V is larger than that for MoN nanoparticles (0.670) (Fig. 4.4a,b). Synergistic effect was the reason for improving the photovoltaic performance of MoN. Figure 4.4(c, d) reveals that the addition of CNTs enhanced electrocatalytic activity and ionic conductivity.

4.4 Titanium Nitride (TiN)

Rather than VN and MoN, Guanglei Cui et al. synthesized three different size of TiNs micro-/nanorods (namely TiN-1, TiN-2, and TiN-3) and are employed as a CE in DSSC [45]. The SEM and DSSC performance are displayed in Fig. 4.5. Among all, TiN-1 is found to be best one that provides the performance of V_{OC}=0.741 V, J_{SC}=14.45 mA/cm^2, FF=0.62, and PCE=6.68%, which is close to that in DSSC with Pt-based electrode. Further evaluated the internal electrochemical characteristics to analyze the R_S and R_{CT} at Pt/FTO interface (Fig. 4.5b). The obtained results reveal, R_S and R_{CT} value of the TiN-based CEs is much lower than that of the others which imitates tuning of the electrode size can further optimize the solar PCE. Further, importance of organic materials with TiN CE-based DSSCs are tabulated in Table 4.3.

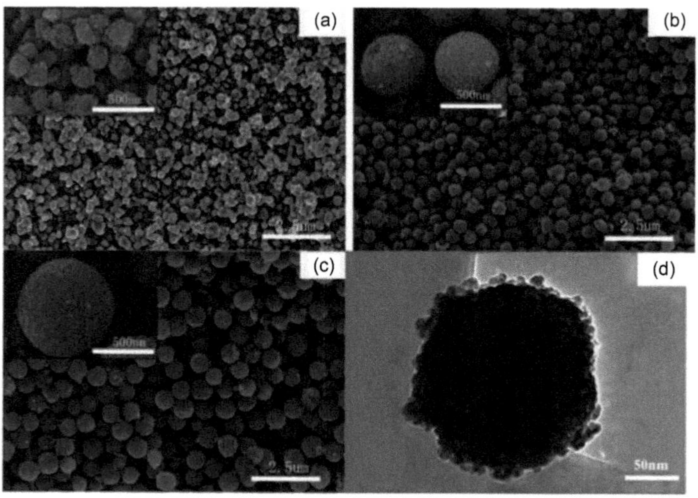

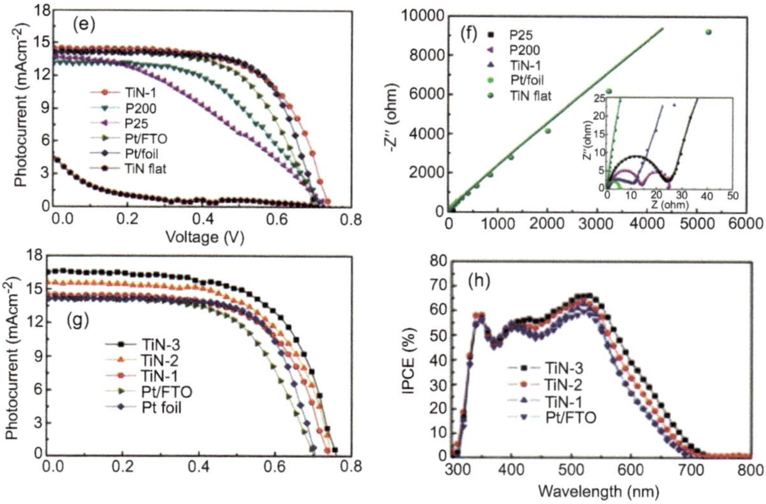

Figure 4.5 SEM image of the various hierarchical micro-/nano-TiNs CE films (a) TiN-1, (b) TiN-2, and (c) TiN-3. (d) TEM image of TiN-1. (e, f) *J–V* and Nyquist curves of the DSSC with different CEs. (g, h) IPCE spectrum. Reprinted with the permission from Ref. [45], Copyright 2012, The Royal Society of Chemistry.

Table 4.3 Performance of DSSC with TN-based CEs

Counter electrode	Sensitizer	Photo-anode	J_{SC} (mA/cm^2)	V_{OC} (V)	FF	PCE (%)	Ref.
TiN NPs	N719	TiO_2	9.28	0.66	0.35	2.12	[46]
CNT			9.55	0.70	0.52	3.53	
TiN-CNTs			12.74	0.75	0.57	5.41	
Pt			12.83	0.73	0.60	5.68	
TiN (R)	N719	TiO_2	13.49	0.78	0.51	5.43	[47]
TiN (R). PEDOT:PSS			14.53	0.72	0.65	6.89	
PEDOT:PSS			10.09	0.71	0.28	2.04	
Pt			13.09	0.74	0.67	6.57	
OM TiN-C	N719	TiO_2	14.36	0.69	0.67	6.71	[48]
OMC			14.48	0.66	0.54	5.16	
TIN NPs			14.08	0.68	0.18	1.73	
Pt			15.44	0.65	0.33	3.32	

4.5 Nickel Nitride

Recently D. Devaraj et al. [49] fabricated DSSC with Ni_3N CEs in E1- and E2-based ionic liquid electrolytes (Fig. 4.6a).

E1: DMPII (0.5M)–LiI (0.1M)–I_2 (0.05M)–GuNC (0.1 M)–4-tBP (0.5 M) in methoxy propionitrile (MPN);

E2: DMPII (0.5M)–LiI (0.1M)–I_2 (0.05M)–GuNC (0.1M)–4-tBP (0.5 M) in acetonitrile (AcN).

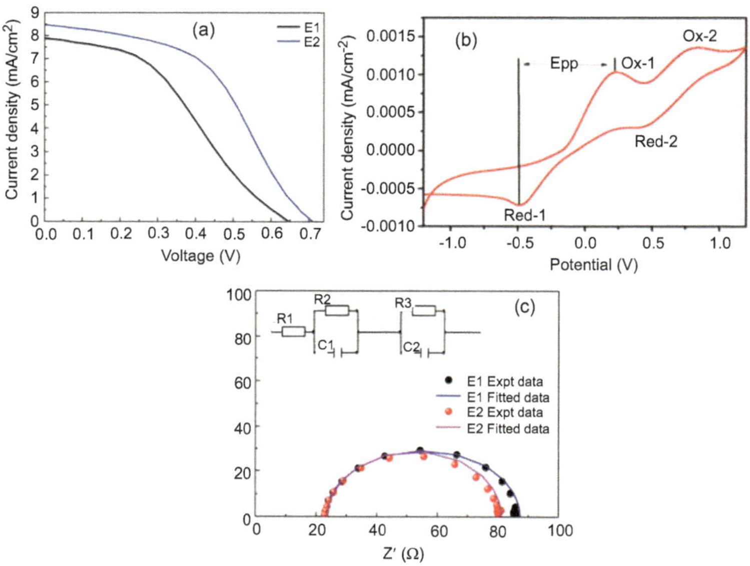

Figure 4.6 (a) *J–V* characteristic performance of E1 and E2. (b) Cyclic voltammogram of the Ni_3N CE. (c) EIS study for E1- and E2-based cells. Reprinted with the permission from Ref. [49], Copyright 2018, The Royal Society of Chemistry.

The DSSC with methoxy propionitrile (MPN)-based electrolyte (E1) exhibits a PCE of 1.96% as compared to the one with acetonitrile (AcN) electrolyte ~2.88%. The low FF and poor electrocatalytic activity of the Ni_3N CEs in electrolyte was the reason for the low power conversion performance (Fig. 4.6b). At the same time, J_{SC} of the device was higher in E1 electrolyte which is mainly due to volatility of the solvent. J_{SC}, FF, and PCE of DSSC with E1 is increased compared to the one with E2 which may be due to higher migration and penetration of E1 into porous TiO_2. The Nyquist plots and the equivalent circuit models were fitted, comprising R_S and R_{CT} (Fig. 4.6c). The electrolyte E2 shows a lower resistance which reflects the device performance. The peak current density and E_{PP} are two major factors for assessing the electrocatalytic performances of CE. A higher I_{PC} indicates faster reaction rate whereas a lower E_{PP} shows a smaller over-potential and improves electrocatalytic behavior for I^-/I_3^- redox reactions. Hence, the higher I_{PC} and a lower E_{PP} are advantageous for a Ni_3N CE to achieve good electrocatalytic activity.

4.6 Zinc Nitride

The DSSCs employed first time using Zn-compounds-based CEs of the Zn_3N_2/PEDOT:PSS 10wt% (Zn_3N_2-10), ZnO/PEDOT:PSS 10wt% (Zn_3N_2-10), ZnS/PEDOT:PSS 10wt% (Zn_3N_2-10), ZnSe/PEDOT:PSS 10wt% (Zn_3N_2-10), bare PEDOT:PSS, and Pt-based films are shown in Fig. 4.7a. Initially, bare Pt and PEDOT:PSS were used as a CE and among PEDOT:PSS cell delivered poor values of J_{SC} and FF owing to the insufficient electrocatalytic active sites while the low V_{OC} obtained due to the poor electrocatalytic activcity (Fig. 4.7b). The addition of PEDOT:PSS provides best performance than the ones without PEDOT:PSS due to fast and continuous conducting matrix and tight adhesion with transparent conducting substrate. Thus,

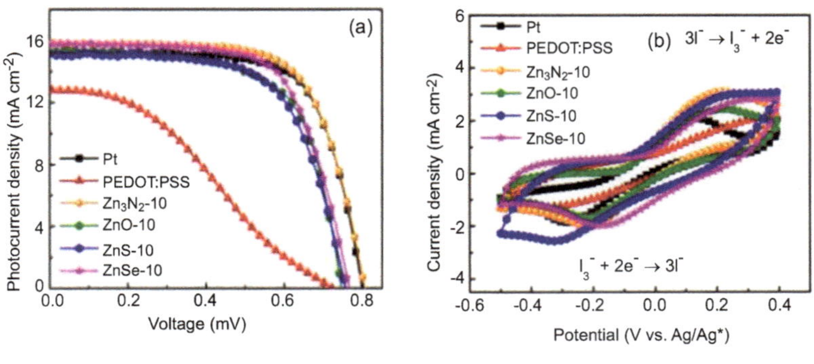

Figure 4.7 (a) JV and (b) CV curve for various CEs respectively. Reprinted with the permission from Ref. [50], Copyright 2015, American Chemical Society.

among other nitrides, Zn_3N_2-10 shows the best PCE of 8.73%, with a V_{OC} of 0.81 V, J_{SC} of 15.77 mA/cm^2, and FF of 0.69. The observed V_{OC}, J_{SC}, and FF of different Zn-based CEs show the tendency of Zn_3N_2-10 > ZnSe-10 > ZnO-10 > ZnS-10, accordingly the same tendency for the values of PCE. In summary, the composite film with Zn_3N_2-10 exhibits the best electrocatalytic ability is confirmed [50].

4.7 Tantalum Nitride

Sining Yun et al. reported the DSSCs with employed three Ta-based binary compound CEs (TaOx, Ta_3N_5, and Ta_4C_3) [51]. The Ta-based CEs resulted in a good photovoltaic performance, with high PCEs of 5.03, 6.79, 7.39, and 7.57% for Ta_3N_5, TaOx, Ta_4C_3, and Pt, respectively (Fig. 4.8). The charge transfer resistance at the interface and electrocatalytic studies are carried out and the results reveal Ta-based CEs to replace Pt in future.

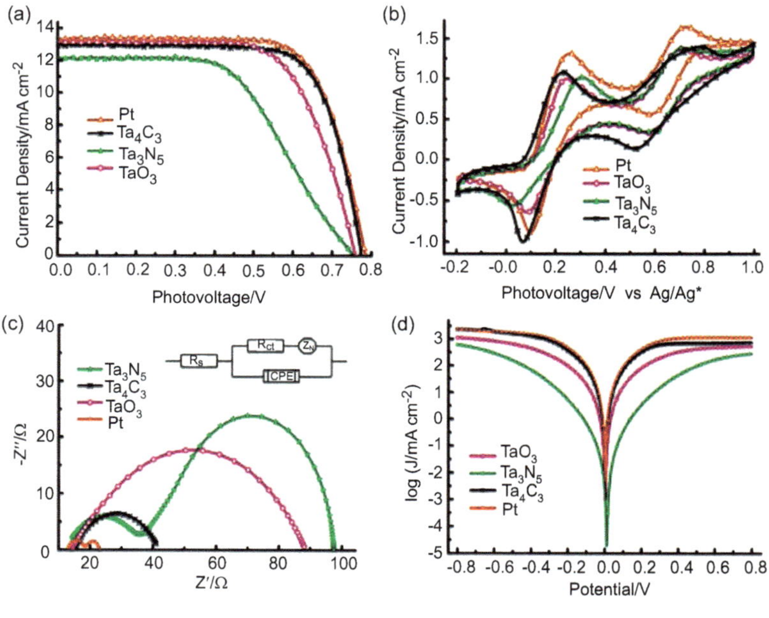

Figure 4.8 (a, b) *J–V* and CVs curves for DSSCs using different CEs under I/I_3 redox couple. (c) Nyquist plot. (d) Tafel plot for different CEs. Reprinted with the permission from Ref. [51], Copyright 2013, John Wiley and Sons.

4.8 Conclusion

The noble metal platinum (Pt) is commonly used as a CE for DSSCs because of its high catalytic activity, high electrical conductivity, and good chemical stability. But, the high cost and low availability of Pt restrict its mass application in DSSCs. Therefore, developing low-cost Pt-free CE material has become an increasingly popular research topic. Carbon materials and conductive polymers have been suggested for CE materials. Recently, nitrides have successfully been introduced into DSSCs, because of some of their advantages over Pt as follows:

i. In general, thermal evaporation technique is used to prepare Pt electrode whereas simple chemical route is enough for TMN-based electrodes.
ii. One can tune the size and morphology of the TMN-based nanomaterials during chemical synthesis which in turn tune the electrocatalytic behavior of the counter electrode.
iii. They are of low cost and are easily available.

The transition metal nitrides displayed superior electrocatalytic activity for reducing I_3^- to I^- and due to their noble metal-like surface electronic structure and the energy level. But it is not easy to get good electrical conductivity and superior electrocatalytic activity simultaneously even though it is an important requirement for good CEs in DSSCs. In the form of nanostructured metal nitride materials provide superior electrocatalytic activity and also enhance the electron transport barriers because of abundant particle boundaries. Replacing Pt CE by metal nitrides has displayed comparable solar cell performance with good electrocatalytic activity. Using this review, we concluded that the titanium-based counter electrodes show higher solar cell efficiency (6.68%) than Pt (6.04%). These enhancements mainly depend on the surface morphology, changes in particle size, and higher surface area due to the various synthesis methods. Further, titanium-based counter electrode is nontoxic and cost-wise cheaper than the platinum as shown in Table 4.4.

Table 4.4 Comparison of cost, toxicity, and availability of nitride-based CE with that of platinum

S. No.	Compound name	Vanadium nitride	Titanium nitride	Molybdenum Nitride	Tantalum nitride	Zinc Nitride	Platinum
1	**Cost (in USD)**	34401 per kg	2670.98 per kg	-	397.01 per litre	-	36836.63 per kg
2	**Toxicity**	Yes	No	Yes	No	No	No
3	**Availability**	✓	✓	X	✓	X	✓

Acknowledgment

I. John Peter thanks the Department of Science and Technology, India for the financial support under DST-INSPIRE fellowship scheme (Code: IF160974). CR thanks GRI-DTBU for the financial support through seed money grant. SZK acknowledges INTPART Funding for International Partnerships project 309827 from the Research Council of Norway.

References

1. R. A. Ezekowitz, D. J. Williams, H. Koziel, M. Y. Armstrong, A. Warner, F. F. Richards, Uptake of *Pneumocystis carinii* mediated by the macrophage mannose receptor, *Nature*, 351 (1991) 155.
2. K. Kakiage, Y. Aoyama, T. Yano, K. Oya, J. I. Fujisawa, M. Hanaya, Highly-efficient dye-sensitized solar cells with collaborative sensitization by silyl-anchor and carboxy-anchor dyes, *Chem. Commun.*, 51 (2015) 15894–15897. doi:10.1039/c5cc06759f.
3. A. Kongkanand, K. Tvrdy, K. Takechi, M. Kuno, P. V. Kamat, N. Dame, Quantum dot solar cells. Tuning photoresponse through size and shape control of CdSe–TiO_2 architecture, *J. Am. Chem. Soc.*, 130 (2008) 4007–4015.
4. R. Vogel, P. Hoyer, H. Weller, Quantum-sized PbS, CdS, Ag_2S, Sb_2S_3, and Bi_2S_3 particles as sensitizers for various nanoporous wide-bandgap semiconductors, *J. Phys. Chem.*, 98 (1994) 3183–3188. doi:10.1021/j100063a022.
5. Y. Yang, W. Rodríguez-Córdoba, T. Lian, Multiple exciton generation and dissociation in PbS quantum dot-electron acceptor complexes, *Nano Lett.*, 12 (2012) 4235–4241. doi:10.1021/nl301847r.
6. T. Shen, J. Tian, L. Lv, C. Fei, Y. Wang, T. Pullerits, G. Cao, Investigation of the role of Mn dopant in CdS quantum dot sensitized solar cell, *Electrochim. Acta.*, 191 (2016) 62–69. doi:10.1016/j.electacta.2016.01.056.
7. D. Katz, T. Wizansky, O. Millo, E. Rothenberg, T. Mokari, U. Banin, Size-dependent tunneling and optical spectroscopy of CdSe quantum rods, *Phys. Rev. Lett.*, 89 (2002) 1–4. doi:10.1103/PhysRevLett.89.086801.
8. Z. Li, C. Dong, L. Tang, X. Zhu, H. Chen, J. Ren, Aqueous synthesis of CdTe/CdS/ZnS quantum dots and their optical and chemical properties, *Luminescence*, 26 (2011) 439–448. doi:10.1002/bio.1250.

9. M.M. Tavakoli, H. Aashuri, A. Simchi, S. Kalytchuk, Z. Fan, Quasi core/shell lead sulfide/graphene quantum dots for bulk heterojunction solar cells, *J. Phys. Chem. C* 119 (2015) 18886–18895. doi:10.1021/acs.jpcc.5b04195.
10. A. J. Nozik, Quantum dot solar cells, *Phys. E Low Dimens. Syst. Nanostruct.*, 14 (2002) 115–120. doi:10.1016/S1386-9477(02)00374-0.
11. I. J. Peter, K. Ramachandran, S. Vijaya, S. Anandan, P. Nithiananthi, Effect of phosphor on the efficiency of TiO_2/CdS/Ag_2S heterostructure based solar cells, *Mater. Lett.*, 240 (2019) 291–294. doi:10.1016/j.matlet.2019.01.031.
12. A. Tubtimtae, M. W. Lee, G. J. Wang, Ag_2Se quantum-dot sensitized solar cells for full solar spectrum light harvesting, *J. Power Sources*, 196 (2011) 6603–6608. doi:10.1016/j.jpowsour.2011.03.074.
13. I. Zumeta-Dubé, V.-F. Ruiz-Ruiz, D. Díaz, S. Rodil-Posadas, A. Zeinert, TiO_2 sensitization with Bi_2S_3 quantum dots: The inconvenience of sodium ions in the deposition procedure, *J. Phys. Chem. C*, 118 (2014) 11495–11504.
14. T. L. Li, Y. L. Lee, H. Teng, High-performance quantum dot-sensitized solar cells based on sensitization with $CuInS_2$ quantum dots/CdS heterostructure, *Energy Environ. Sci.*, 5 (2012) 5315–5324. doi:10.1039/c1ee02253a.
15. J. Du, Z. Du, J. S. Hu, Z. Pan, Q. Shen, J. Sun, D. Long, H. Dong, L. Sun, X. Zhong, L. J. Wan, Zn–Cu–In–Se quantum dot solar cells with a certified power conversion efficiency of 11.6%, *J. Am. Chem. Soc.*, 138 (2016) 4201–4209. doi:10.1021/jacs.6b00615.
16. I. John Peter, N. Rajamanickam, S. Vijaya, S. Anandan, K. Ramachandran, P. Nithiananthi, TiO_2/Graphene quantum dots core-shell based photo anodes with TTIP treatment: A perspective way of enhancing the short circuit current, *Sol. Energy Mater. Sol. Cells* 205 (2020) 110239. doi: 10.1016/j.solmat.2019.110239.
17. V. Gayathri, I. J. Peter, N. Rajamanickam, K. Jayakumar, K. Ramachandran, Influence of Li doping on the conversion efficiency of nano TiO_2 based solar cell, *AIP Conf. Proc.*, 2115 (2019) 030544. doi:10.1063/1.5113383.
18. V. Gayathri, I. John Peter, N. Rajamanickam, K. Ramachandran, Improved performance of dye-sensitized solar cells by Cr doped TiO_2 nanoparticles, *Mater. Today Proc.*, (2019) 2–5. doi:10.1016/j.matpr.2019.05.381.
19. J. Wang, Y. Li, Q. Shen, T. Izuishi, Z. Pan, K. Zhao, X. Zhong, Mn doped quantum dot sensitized solar cells with power conversion efficiency

exceeding 9%, *J. Mater. Chem. A.*, 4 (2016) 877–886. doi:10.1039/c5ta09306f.

20. Y. L. Lee, C. H. Chang, Efficient polysulfide electrolyte for CdS quantum dot-sensitized solar cells, *J. Power Sources*, 185 (2008) 584–588. doi:10.1016/j.jpowsour.2008.07.014.
21. Y. Tachibana, K. Umekita, Y. Otsuka, S. Kuwabata, Performance improvement of CdS quantum dots sensitized TiO_2 solar cells by introducing a dense TiO_2 blocking layer, *J. Phys. D. Appl. Phys.*, 41 (2008) 102002. doi:10.1088/0022-3727/41/10/102002.
22. R. M. Evangelista, S. Makuta, S. Yonezu, J. Andrews, Y. Tachibana, Semiconductor quantum dot sensitized solar cells based on ferricyanide/ferrocyanide redox electrolyte reaching an open circuit photovoltage of 0.8 V, *ACS Appl. Mater. Interfaces*, 8 (2016) 13957–13965. doi:10.1021/acsami.6b03633.
23. H. J. Lee, J. H. Yum, H. C. Leventis, S. M. Zakeeruddin, S. A. Haque, P. Chen, S. Il Seok, M. Grätzel, M. K. Nazeeruddin, CdSe quantum dot-sensitized solar cells exceeding efficiency 1% at full-sun intensity, *J. Phys. Chem. C*, 112 (2008) 11600–11608. doi:10.1021/jp802572b.
24. L. Li, X. Yang, J. Gao, H. Tian, J. Zhao, A. Hagfeldt, L. Sun, Highly efficient CdS quantum dot-sensitized solar cells based on a modified polysulfide electrolyte, *J. Am. Chem. Soc.*, 133 (2011) 8458–8460. doi:10.1021/ja201841p.
25. J. Wang, Y. Li, Q. Shen, T. Izuishi, Z. Pan, K. Zhao, X. Zhong, Mn doped quantum dot sensitized solar cells with power conversion efficiency exceeding 9%, *J. Mater. Chem. A*, 4 (2016) 877–886. doi:10.1039/c5ta09306f.
26. J. G. Radich, R. Dwyer, P. V Kamat, Cu_2S reduced graphene oxide composite for high-efficiency quantum dot solar cells. Overcoming the redox limitations of S_2^-/S_n^{2-} at the counter electrode, *J. Phys. Chem. Lett.*, 2 (2011) 2453–2460.
27. Z. Tachan, M. Shalom, I. Hod, S. Rühle, S. Tirosh, A. Zaban, PbS as a highly catalytic counter electrode for polysulfide-based quantum dot solar cells, *J. Phys. Chem. C*, 115 (2011) 6162–6166. doi:10.1021/jp112010m.
28. S. Wang, J. Tian, Recent advances in counter electrodes of quantum dot-sensitized solar cells, *RSC Adv.*, 6 (2016) 90082–90099. doi:10.1039/c6ra19226b.
29. K. Meng, G. Chen, K. R. Thampi, Metal chalcogenides as counter electrode materials in quantum dot sensitized solar cells: A perspective, *J. Mater. Chem. A*, 46 (2015). doi:10.1039/c5ta05071e.

30. V. T. Chebrolu, H. J. Kim, Recent progress in quantum dot sensitized solar cells: An inclusive review of photoanode, sensitizer, electrolyte, and the counter electrode, *J. Mater. Chem. C*, 7 (2019) 4911–4933. doi:10.1039/c8tc06476h.

31. C. Gao, Q. Han, M. Wu, Review on transition metal compounds based counter electrode for dye-sensitized solar cells, *J. Energy Chem.*, 27 (2018) 703–712. doi:10.1016/j.jechem.2017.09.003.

32. J. Theerthagiri, A. R. Senthil, J. Madhavan, T. Maiyalagan, Recent progress in non-platinum counter electrode materials for dye-sensitized solar cells, *ChemElectroChem*, 2 (2015) 928–945. doi:10.1002/celc.201402406.

33. K. M. Bratlie, H. Lee, K. Komvopoulos, P. Yang, G. A. Somorjai, Platinum nanoparticle shape effects on benzene hydrogenation selectivity, *Nano Lett.*, 7 (2007) 3097–3101. doi:10.1021/nl0716000.

34. G. R. Li, X. P. Gao, Low-cost counter-electrode materials for dye-sensitized and perovskite solar cells, *Adv. Mater.*, 32 (2020) 1–20. doi:10.1002/adma.201806478.

35. H. T. Chen, Y. J. Huang, C. T. Li, C. P. Lee, J. T. Lin, K. C. Ho, Boron nitride/sulfonated polythiophene composite electrocatalyst as the TCO and Pt-free counter electrode for dye-sensitized solar cells: 21% at dim light, *ACS Sustain. Chem. Eng.*, 8 (2020) 5251–5259. doi:10.1021/acssuschemeng.0c00097.

36. M. Wu, H. Guo, Y.N. Lin, K. Wu, T. Ma, A. Hagfeldt, Synthesis of highly effective vanadium nitride (VN) peas as a counter electrode catalyst in dye-sensitized solar cells, *J. Phys. Chem. C*, 118 (2014) 12625–12631. doi:10.1021/jp501797e.

37. G. Wang, S. Hou, C. Yan, W. Zhang, A 3D architecture composite of porous vanadium nitride nanoribbons and reduced graphene oxide as a high-efficiency counter electrode for dye-sensitized solar cells, *RSC Adv.*, 8 (2018) 1083–1088. doi:10.1039/c7ra11279c.

38. P. Vijayakumar, M. Senthil Pandian, A. Pandikumar, P. Ramasamy, Electrochemical interfacial charge transfer dynamics and photovoltaic performances of nanofibrous vanadium derivatives based platinum free counter electrodes in dye sensitized solar cells, *Mater. Sci. Eng. B*, 222 (2017) 7–17. doi:10.1016/j.mseb.2017.04.005.

39. X. Zhang, X. Chen, K. Zhang, S. Pang, X. Zhou, H. Xu, S. Dong, P. Han, Z. Zhang, C. Zhang, G. Cui, Transition-metal nitride nanoparticles embedded in N-doped reduced graphene oxide: Superior synergistic electrocatalytic materials for the counter electrodes of dye-sensitized

solar cells, *J. Mater. Chem. A*, 1 (2013) 3340–3346. doi:10.1039/c2ta00608a.

40. M. Wu, X. Lin, Y. Wang, L. Wang, W. Guo, D. Qi, X. Peng, A. Hagfeldt, M. Grätzel, T. Ma, Economical Pt-free catalysts for counter electrodes of dye-sensitized solar cells, *J. Am. Chem. Soc.*, 134 (2012) 3419–3428. doi:10.1021/ja209657v.
41. J. Song, G. R. Li, K. Xi, B. Lei, X. P. Gao, R. V. Kumar, Enhancement of diffusion kinetics in porous MoN nanorods-based counter electrode in a dye-sensitized solar cell, *J. Mater. Chem. A*, 2 (2014) 10041–10047. doi:10.1039/c4ta01342e.
42. G. R. Li, J. Song, G. L. Pan, X. P. Gao, Highly Pt-like electrocatalytic activity of transition metal nitrides for dye-sensitized solar cells, *Energy Environ. Sci.*, 4 (2011) 1680–1683. doi:10.1039/c1ee01105g.
43. M. Wu, Q. Zhang, J. Xiao, C. Ma, X. Lin, C. Miao, Y. He, Y. Gao, A. Hagfeldt, T. Ma, Two flexible counter electrodes based on molybdenum and tungsten nitrides for dye-sensitized solar cells, *J. Mater. Chem.*, 21 (2011). doi:10.1039/c1jm11422k.
44. J. Song, G. R. Li, F. Y. Xiong, X. P. Gao, Synergistic effect of molybdenum nitride and carbon nanotubes on electrocatalysis for dye-sensitized solar cells, *J. Mater. Chem.*, 22 (2012) 20580–20585. doi:10.1039/c2jm34878k.
45. X. Zhang, X. Chen, S. Dong, Z. Liu, X. Zhou, J. Yao, S. Pang, H. Xu, Z. Zhang, L. Li, G. Cui, Hierarchical micro/nano-structured titanium nitride spheres as a high-performance counter electrode for a dye-sensitized solar cell, *J. Mater. Chem.*, 22 (2012) 6067–6071. doi:10.1039/c2jm30420a.
46. G. R. Li, F. Wang, Q. W. Jiang, X. P. Gao, P. W. Shen, Carbon nanotubes with titanium nitride as a low-cost counter–electrode material for dye-sensitized solar cells, *Angew. Chem. Int. Ed. Engl.*, 49 (2010) 3653–3656. doi:10.1002/anie.201000659.
47. H. Xu, X. Zhang, C. Zhang, Z. Liu, X. Zhou, S. Pang, X. Chen, S. Dong, Z. Zhang, L. Zhang, P. Han, X. Wang, G. Cui, Nanostructured titanium nitride/PEDOT:PSS composite films as counter electrodes of dye-sensitized solar cells, *ACS Appl. Mater. Interfaces*, 4 (2012) 1087–1092. doi:10.1021/am201720p.
48. E. Ramasamy, C. Jo, A. Anthonysamy, I. Jeong, J. K. Kim, J. Lee, Soft-template simple synthesis of ordered mesoporous titanium nitride-carbon nanocomposite for high performance dye-sensitized solar cell counter electrodes, *Chem. Mater.*, 24 (2012) 1575–1582. doi:10.1021/cm203672g.

49. S. Prasad, G. Durai, D. Devaraj, M. S. AlSalhi, J. Theerthagiri, P. Arunachalam, M. Gurulakshmi, M. Raghavender, P. Kuppusami, 3D nanorhombus nickel nitride as stable and cost-effective counter electrodes for dye-sensitized solar cells and supercapacitor applications, *RSC Adv.*, 8 (2018). doi:10.1039/c8ra00347e.

50. C. T. Li, H. Y. Chang, Y. Y. Li, Y. J. Huang, Y. L. Tsai, R. Vittal, Y. J. Sheng, K. C. Ho, Electrocatalytic zinc composites as the efficient counter electrodes of dye-sensitized solar cells: Study on the electrochemical performances and density functional theory calculations, *ACS Appl. Mater. Interfaces*, 7 (2015). doi:10.1021/acsami.5b07724.

51. S. Yun, M. Wu, Y. Wang, J. Shi, X. Lin, A. Hagfeldt, T. Ma, Pt-like behavior of high-performance counter electrodes prepared from binary tantalum compounds showing high electrocatalytic activity for dye-sensitized solar cells, *ChemSusChem*, 6 (2013) 411–416. doi:10.1002/cssc.201200845.

Chapter 5

Potential Development of Transition Metal Sulphides Based Counter Electrode Platform for Dye-Sensitized Solar Cells

K. S. Rajni and T. Raguram
Department of Sciences, Amrita School of Engineering,
Amrita Vishwa Vidyapeetham, Coimbatore 641112,
Tamil Nadu, India
ks_rajani@cb.amrita.edu

5.1 Introduction

Dye-sensitized solar cell (DSSC) pops out to be the most prominent third-generation solar cell, challenging the conventional silicon-based solar cells in various aspects, such as semi-transparency, fabrication, and environmental friendliness. The power provided by the sun in one minute can meet the energy needs of the world for 1 year. In a day, it provides us more energy than that would be consumed by our current population in 27 years. Also, all the energy stored in the fossil fuels is equivalent to the energy received from the sun for three days. Even though solar energy is inexhaustible and

Counter Electrode for Dye-Sensitized Solar Cells
Edited by Alagarsamy Pandikumar, Ramesh Mohan, and Kandasamy Jothivnekatachalam

ISBN 978-981-4877-38-1 (Hardcover), 978-1-003-11077-4 (eBook)
www.jennystanford.com

free, trapping and utilizing it is a challenging task. DSSC also has a number of distinctive characteristics, such as simple to fabricate using the conventional techniques, semi-flexible and semi-transparent, and offers a wide variety of uses. DSSC consists of mesoporous TiO_2 (anatase) photoanode, sensitized by dye molecules (organic/inorganic), and a platinum (Pt) counter electrode (CE) sandwiched by liquid electrolyte. Photoanodes play two roles, one as the support for the sensitizer loading and the other as the transporter of photoexcited electrons to the external circuit from the sensitizer. The absorption of sunlight and conversion of solar energy into electrical energy is carried out by dye molecules. Electrolyte acts as a mediator to provide internal pathways for transporting electrons from cathode to anode. The function of CE is to collect electrons from external circuit and promotes the regeneration of I^- from I_3^- ions.

An excellent CE material should possess perfect stability and high catalytic activity. Since Pt possesses these two properties, it has been widely employed counter electrode material for DSSC. As Pt is an expensive noble metal, its use in DSSC has not been commercialized. Also, the heat treatments during the preparation of the conventional Pt CEs exclude the choice of a flexible plastic substrate. These reasons have led to the development of inexpensive new materials for counter electrode in DSSC. Several materials, such as carbon nanostructures, activated carbon, carbon nanotubes (CNTs), graphene, and conducting polymers and their various combinations, have been tried as CE materials. In the case of the carbon material, it is noted that the fill factor (FF) improves and the charge transfer resistance (R_{ct}) decreases with increase in the thickness of the CE, indicating the high catalytic activity that results in the improvement of the overall cell performance. Carbon compounds are cost-effective materials compared to Pt and also show excellent electrochemical activity towards the redox species. But the adhesion towards the transparent conducting oxide (TCO) substrate has a long-term stability issue. Transition metal sulfides are one of the promising emerging materials capable of replacing Pt and graphene CEs. Under this category, binary, ternary, quaternary, and penternary sulfides are tried due to their excellent electrocatalytic activity for the reduction of I_3^- and also their plentiful availability and low cost compared to Pt. Their exciting properties, such as durability towards the covalent solid, high melting temperature of the ionic crystal, and electrical and thermal conductivities, make them a most striking one for CE in DSSC. Among the family of metal sulfides, nickel sulfide has attracted much attention due to its potential application in hydrogen evolution and lithium batteries. However, the long-term stability of nickel (Ni) substrate in the corrosive I^-/I_3^- electrolyte of DSSCs is still unknown and the electron transport between the nanoparticles and the conducting substrate is low and this hinders the overall performance of DSSCs. However, this drawback could be overcome by incorporating carbon nanostructures into the transition metal compounds, which will improve the catalytic performance of non-noble metals. Thus, the development and evaluation of new organic and inorganic compounds as an alternative material to Pt in DSSCs becomes a promising research area for minimizing the drawbacks and the cost of devices for commercialization. This chapter focuses on the catalytic activity of the binary, ternary, quaternary, and penternary metal sulfide counter electrodes and the resultant efficiencies of DSSC [1, 2].

5.2 Binary Transition Sulfides

Vanadium sulfide (VS_2): CE prepared by in situ hydrothermal route at different temperatures (140°C, 160°C, 180°C, 200°C) [3] exhibited a conversion efficiency of 6.24% under an illumination of 100 mW/cm^2 as compared to the DSSC based on the Pt CE. Tafel curves for VS_2 CEs prepared at different temperatures show the exchange current density (J_0), obtained as the intercept of the extrapolated linear region of the curve when the overpotential is zero, which is positively correlated to the reduction capability of CE materials in I^-/I_3^- electrolyte. From the Tafel plot, it is noted that VS_2 (180°C preparation) nanofibre structure is better for electron transport than VS_2 nanoparticles. Thus, this indicates that the electrochemical catalytic activity of VS_2 CE is significantly affected by the morphology. From the electrochemical impedance spectroscopy (EIS) analysis, the first semicircle in the high frequency region denotes the R_{ct} at the CE/electrolyte interface [3, 4]. The smaller R_{ct} exhibits a faster electron transfer from the CE to the electrolyte, which is a significant parameter for evaluating the performance of CEs. It is noted that the R_{ct} value for the VS_2 CE prepared at 180°C is comparable to that of the Pt CE. The J–V analysis of Pt- and VS_2 CE- (180°C preparation) based DSSCs, under the irradiation of

100 mW/cm^2, shows the positive effect for VS_2 CEs and this possibly results (efficiency η = 6.44%) from the contact frequency between the I^-/I_3^- redox couple in the electrolyte. Also, the CE can be quickened because of the large specific surface area of VS_2 nanofibers, such that to provide a good reaction speed for the reduction of I^-/I_3^- ions. In addition, VS_2 nanofibers have synergic advantages of high conductivity and 2D permeable channels guarantee the rapid transmission of electrons, thereby showing efficient of photoconversion efficiency (PCE) for the DSSC based on the VS_2 counter electrode.

CNTs/VS_2: A hybrid material of CNT/VS_2 nanofibre CE prepared by the in situ hydrothermal route [3] and the synergetic effect of CNTs (with superior electronic conductivity) and VS_2 (with excellent catalytic ability) show an increase in the photovoltaic property of devices. Figure 5.1 shows the morphologies of the VS_2 and CNTs/VS_2 CE prepared at 180°C (marked VS_2 and CNTs/VS_2, respectively). It shows VS_2 nanofibres with short size and dendritic structures. Such structure facilitates the adsorption of large amounts of electrolyte and provides more electron transport channels. One can observe VS_2 nanoparticles changed to nanofibres when the temperature increases from 140°C to 180°C and then to two-dimensional nanosheets or nano-petal-like structures when the temperature is increased to 200°C. The probable cause is that the supersaturation and crystal growth rate become higher when the temperature increases [3]. The cyclic voltammetry (CV) for various CEs with varying CNT of 0.01, 0.03, 0.05, 0.07, and 0.09 g show two pairs of redox peaks in the range from –0.4 to 0.4 potential, which has a significant impact on the electrochemical catalytic properties of CE in DSSCs. The cathodic peak current density (J_{pc}) is positive and it is correlated with the reaction rate of the CE for the reduction of I_3^- ions to I^- ions [5, 6]. The voltammograms of CNTs/VS_2 CEs with different synthesis temperatures (140°C, 160°C, 180°C, and 200°C) have similar cathodic peak potentials (V_{pc}) and J_{pc} indicates the significant influence of temperature on the electrochemical catalytic properties and the crystal form of the materials [7]. It is concluded that the significant enhancements in the electrochemical performance of CNTs/VS_2 CE are possible due to the large catalytic surface area and excellent conductivity of CNTs. Further increasing

CNTs above the optimum level (0.07 g), there is a negative effect on the electrocatalytic activities of CEs, thus contributing to the poor catalytic activity of CNTs [8, 9].

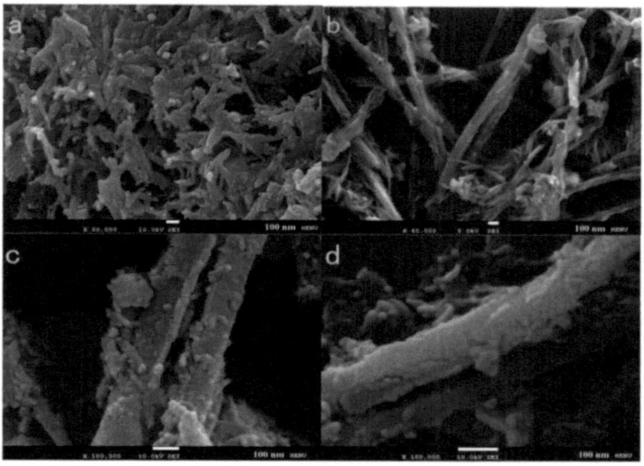

Figure 5.1 SEM images of (a) VS_2 and (b–d) CNTs/VS_2. Reprinted from Ref. [3], Copyright 2017, with permission from Elsevier.

The photovoltaic characteristics of DSSC with CNTs/VS_2 CE show the PCE of 7.72%, open-circuit voltage (V_{oc}) of 0.753 V, short-circuit current density (J_{sc}) of 15.05 mA/cm^2, and FF of 0.681, which are much higher than those of the DSSCs based on VS_2 and Pt CEs. The results show that moderate amounts of CNTs can enhance the photoelectric performance of DSSCs due to the high specific surface area and excellent conductivity [3].

Cobalt sulfide (CoS_2): Mesoporous cobalt sulfide nanotube arrays CE on FTO shows the efficiencies of 2.13%, 5.46%, 6.11%, and 6.13% as reported by J. C. Tsai et al. [10]. The CEs are synthesized by combining three simple technologies, namely, selective etching of ZnO sacrificial templates, mesoporous Co_3O_4 formation from cobalt-chelated chitosan, and increasing the time of ion exchange reaction (IER) [10]. The mesoporous Co_3O_4 nanotubes composed of Co_3O_4 nanoparticles possess a high surface area and are taken advantage for further removal of templates and IER. The CV curve shows two typical pairs of oxidation and reduction peaks similar to that the Pt counter electrode. The relative negative pair corresponds to the redox reaction of I^-/I_3^-, while the positive pair is assigned to the redox reaction of I_2/I_3^- [11, 12].

For four cobalt sulfide electrodes (NT-1 to NT-4), the catalytic ability increases with increasing time of IER and is attributed to the degree of sulfur replacement for oxygen ions during the IER process. The reaction time does not affect significantly the morphological features of cobalt sulfide, which means a higher amount of CoS_2 can be transformed from Co_3O_4 for nanotubes prepared using longer IER times and the electrodes show better electrocatalytic abilities. The cathodic peak potential of the mesoporous CoS_2 nanotube array electrode is 0.06 V, which is more positive than that of the sputtered Pt electrode (0.17 V). The photovoltaic characteristics of DSSCs using different mesoporous CoS_2 nanotube array CEs, as well as the sputtered Pt as a reference, suggest that Co_3O_4 in CoS_2 nanotube arrays results in a reduced catalytic activity and conductivity of the CE, which drastically affect the solar cell performance. The increase in the photovoltaic parameters can be attributed to the lower charge transfer resistance and sheet resistance. The open-circuit voltage is the difference between the Fermi level of the nanocrystalline TiO_2 film and the redox potential of the mediator that is influenced by intrinsic material properties of the CE and also more interfacial active sites of CEs [13, 14]. Compared to the DSSC with a sputtered Pt CE, DSSCs with CoS_2 nanotube array CEs possess an improved V_{oc}. DSSCs with mesoporous CoS_2 nanotube array CEs show a superior FF and lead to enhanced conversion efficiencies. In theory, the high value of the FF is closely attributed to a low series resistance (R_s) of the whole solar cell. R_s comprises the diffusion resistance of I_3^- in electrolytes (RE), the R_{ct} at the interface of CEs and electrolytes, and the TCO substrate resistance (RTCO) [15]. Thus, mesoporous CoS_2 nanotube arrays are considered to have lower R_{ct} and R_s, which result in the superior FF of DSSCs. Lin et al. [16] reported an electrodeposited mesoporous CoS film on FTO coated glass as a CE in DSSCs, in which V_{oc}, J_{sc}, and FF values were found to be 0.71 V, 13.73 mA/cm^2, and 0.64, respectively, which resulted in a PCE of 6.22%. However, different impurities including Co, $Co(OH)_2$, and formamidine disulfide are easily detected in the mesoporous CoS films prepared by potentiodynamic deposition due to the complicated mechanism of the electrodeposition.

Kung et al. [17] also reported the use of CoS acicular nanorod arrays, synthesized using the reprocess, as the CEs, and the highest photovoltaic performance of the DSSC was shown to have a V_{oc} of 0.71 V, J_{sc} of 16.31 mA/cm^2, and FF of 0.66, resulting in an achieved high conversion efficiency of 7.67%. Unfortunately, the thickness of the CoS arrays was difficult to control in the experiment. On the contrary, a long time (~24 h) was required to obtain CoS nanorod arrays, and this easily causes the oxidation on the surface of CoS to form mosaic flower-like nanoflakes of CoSOH, which can restrain the performance of DSSCs [17]. From the electrochemical analyses and the *J–V* (photocurrent density–voltage) performance of DSSCs, the CE composed of a mesoporous CoS_2 nanotube array is found to be a good candidate for I_3^-/I^- based DSSCs. J. Jin et al. reported [18] a self-assembled CoS_2 film prepared via a facile two-step approach, and it showed high electrocatalytic activity on I^-/I_3^- redox reaction. A maximum power conversion efficiency of 6.78% is achieved for the DSSC based on the resultant CoS_2 nanocrystal CE, which is close to that of the cell using Pt CE (7.38%) under the same condition. Moreover, the flexible device based on CoS_2 CE exhibits an efficiency of 6.40%. The results indicate that the self-assembled CoS_2 nanocrystal thin film may replace the expensive Pt CE for DSSC applications, especially for large-scale flexible devices. From *J–V* curves of the DSSCs using CEs based on CoS_2 film, J_{sc} of 14.62 mA/cm^2, V_{oc} of 0.71, FF of 0.64, and η of 6.78% are noted, and they are comparable to the performance of the DSSC based on the conventional Pt CE (J_{sc} = 14.78 mA/cm^2, V_{oc} = 0.72, FF = 0.68, and η = 7.38%). Thus, the self-assembled CoS_2 nanocrystalline thin film CE is a promising alternative to the precious Pt in DSSC. The EIS measurement (Fig. 5.2) shows two semicircles; the high frequency intercept on the real axis represents the series impedance (R_s), contributed mainly from the resistance of contact, bulky electrolyte and electrodes. The left semicircle arises from charge transfer impedance (R_{ct}) at the CE/electrolyte interface and C_μ (chemical capacitance) represents the double layer capacitance. It is found that the R_s value of CoS_2 (34.20 Ω) is slightly higher than that of Pt (27.13 Ω), suggesting that the electrical conductivity of the self-assembled CoS_2 film is similar to Pt CE in DSSCs. The R_{ct} of CoS_2 is 7.21 Ω, illustrating that CoS_2 has a good catalytic activity for I^-/I_3^- redox reaction. The C_μ values of

CoS_2 and Pt are 27.57 and 2.76 μF, respectively, indicating that a high surface area of CoS_2 as compared to Pt is in favor of catalysis.

In the Tafel polarization curves (Fig. 5.2), the current at high potential (J_{lim}) represents the limiting diffusion rate of triiodide on electrode, and the slope of the curve corresponds to the ion exchange rate on the electrode. It is found that the CoS_2 CE exhibits almost the same Tafel curve as the Pt CE and J_{sc} value is very similar to that for Pt CE (6.21 mA/cm^2). Therefore, the obtained CoS_2 thin film CE via self-assembly shows good electrocatalytic activity on the I^-/I_3^- redox reaction, which is comparable to Pt CE. Being an alternative low-cost CE for Pt electrode, CoS_2 nanoink also has another advantage that it can be printed onto various flexible substrates for large-area roll-to-roll mass production. The device with CoS_2 nanoink onto a flexible ITO/PET substrate exhibits η of 6.40%, J_{sc} of 13.17 mA/cm^2, V_{oc} of 0.70 V, and FF of 0.68 (under AM 1.5 illumination at 98 mW/cm^2). The slight decrease in η for a flexible device compared with the above rigid devices is mainly caused by the increased resistivity that gives rise to a lower current [18].

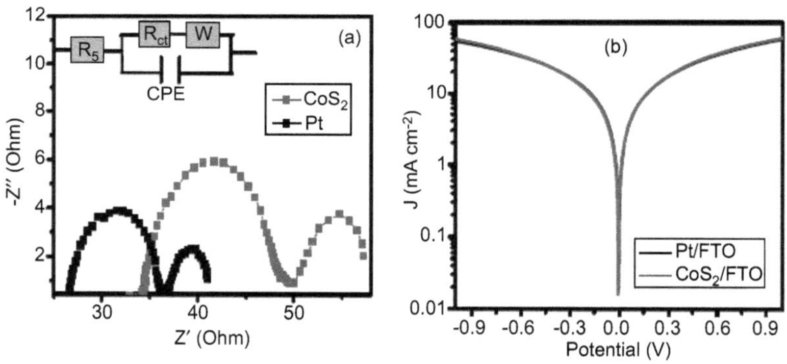

Figure 5.2 Nyquist plots for symmetric cells fabricated with CoS_2 and Pt CE; the inset gives the equivalent circuit and Tafel polarization curves at a scan rate of 50 mV/s with symmetrical cells based on CoS_2 and Pt CE. Reprinted from Ref. [18], Copyright 2014, with permission from American Chemical Society.

Tin sulfide (SnS_2): A new type of semi-transparent SnS_2 nanosheet (NS) [19] CE with 300 nm thickness exhibits excellent electrochemical catalytic activity and shows a conversion efficiency of 7.64% compared with Pt CE (7.71%). Tafel polarization curve for

SnS_2 400°C is similar to that of Pt and shows high stability during the test conditions. Considering the high performance and excellent stability of SnS_2 400°C CEs, SnS_2 nanosheets are expected to have a great potential in replacing the expensive Pt in DSSCs as a promising low-cost, effective, and stable CE catalyst. From the EIS analysis, it is noted that the surface composition, morphology, crystal structure, particle size, and so on could affect the electrochemical properties of the material, and they still maintain a high catalytic activity even though some of the electrode materials have high Nernst diffusion impedance [20–24]. The photovoltaic parameters show an efficiency of 6.30% compared to 6.67% for Pt, and it is attributed to the large reactive surface area due to 5–10 nm sized particles.

Nickel sulphide (NiS_2): Electrocatalytic NiS_2 cube CE synthesized onto FTO using a hydrothermal approach displays an excellent electrocatalytic activity in the reduction of I_3^- and shows a power conversion efficiency of 5.56% compared to Pt-containing DSSC (7.05%) [25]. NiS_2 cubes show a comparable stability to Pt, which is ascribed to the low resistance to diffusion and transfer of electrolyte ions. From EIS experiments, the R_{ct} value of NiS_2 CE is found to be 5.23 Ω, which is higher than that of the Pt CE (3.08 Ω). The structure of NiS_2 cubes endows them with a huge surface area, and thus NiS_2 cubes provide more active sites for the reaction compared to Pt. From the results of Tafel polarization measurement [26], anodic and cathodic branches of NiS_2 cubes show smaller slopes than that of Pt, which indicates that the current density (J_0) of NiS_2 cubes is smaller than that of Pt. In contrast, R_{ct} for the as-prepared NiS_2 CE is greater than that for the Pt CE. The diffusivity of the redox couple and the catalytic characteristics of the CE are determined by limiting current density (J_{lim}) [27]. The J_{lim} of NiS_2 cubes is similar to that of Pt, which means that the redox couple at NiS_2 cubes has a high diffusion coefficient, and reflects the high diffusion rate of the redox couple in the electrolyte. This means that DSSCs containing NiS_2 CEs are expected to have relatively high photovoltaic performances. The calculated photoconversion efficiencies of NiS_2 both cube and octahedron shows that 5.43% and 5.98%, respectively [28]. From the EIS analysis, R_s of NiS_2 octahedron CE is found to be 13.14 Ωcm^2, which is smaller than that of the cubes (14.98 Ωcm^2), and this revealed a better electrical conductivity of the octahedron CE [29].

The R_{ct} of NiS_2 octahedron CE (9.86 Ωcm^2) is smaller than that of cubes (13.17 Ωcm^2), revealing that the octahedrons with (111) facets have better catalytic activity than cubes with (100) facets [30]. The observed low frequency corresponds to the Nernst diffusion impedance (Z_N) of I_3^-/I^- redox couple in the electrolyte [31]. Constant phase element (CPE) values are obtained from the EIS spectrum and NiS_2 octahedron CE exhibited a larger capacitance (39.15 μF) than that of NiS_2 cube CE (29.44 μF), indicating a higher surface area of electrodes, which is responsible for higher PCE [32]. The resistance results revealed that octahedrons had better electrical conductivity and catalytic properties compared to cubes.

Molybdenum sulfide (MoS_2): Figure 5.3 shows the FESEM and TEM images of MoS_2 counter electrodes reported by S. I. Raj et al. [33]. From the figure, it is noted that the vertical films create active edge sites, which increases the catalytic property [34]. The MoS_2 films synthesized at the optimized condition are superior to other parameters due to the change in morphology of FTO films. The increase in temperature above 600°C leads to the damage of FTO substrate (Fig. 5.3), and hence the maximum temperature that FTO can withstand is 600°C. In TEM image and its SEAD pattern (Fig. 5.3C,D), layered films with 0.27 nm (d002) and 0.18 nm (d100) spacing are observed. To estimate the catalytic activities of Pt and MoS_2 CEs for I^-/I_3^- redox couple, cyclic voltammetry (CV) measurements are performed in a three-electrode system using SHE as reference electrode, Pt wire as counter electrode, and the prepared sample as working electrode. A typical CV curve shows two redox peaks that correspond to the reduction of I_3^- (cathodic peak) and the oxidation of I^- (anodic peak), where their voltages are represented by V_{pc} and V_{pa}, respectively. The anodic and cathodic peaks are the key features to study the catalytic activity because they are in charge of the reduction of I_3^- to I^- [35]. The value of peak to peak separation (V_{pp}) (between anodic and cathodic peaks) is negatively correlated with the standard electrochemical rate constant of a redox reaction, while the cathodic current density (I_{pc}) is directly correlated with the electrocatalytic activity of CE [36]. V_{pp} is less in case of MoS_2 than Pt, indicating that electrocatalytic activity of MoS_2 is better than the case of Pt. This is mainly due to the presence of active edge sites of MoS_2 films which enhance the catalytic property of the CE [37].

From the Nyquist plot, the high frequency intercept on the real axis represents R_s. The high frequency region (semicircle) corresponds to the R_{ct} and the corresponding chemical capacitance at the electrode/electrolyte interface. The low frequency region (right semicircle) represents the Z_N of I^-/I_3^- redox couple in the electrolyte [38]. The observed R_s value of Pt is smaller, which indicates that Pt has better conductivity than MoS_2. In contrast, R_{ct} of MoS_2 is lower, indicating better electrocatalytic activity for MoS_2 than Pt due to vertically inclined MoS_2 film on the FTO, which creates active edge sites. Also, we can observe that R_{ct} and R_s values for 50 and 150 sccm CEs are similar, indicating the presence of active edge sites on both CEs. But reflectivity of MoS_2 prepared at 50 sccm is higher due to its crystalline property and higher angle of inclination, which lead to more absorption photons. This makes the MoS_2 CE at 50 sccm superior to other CEs.

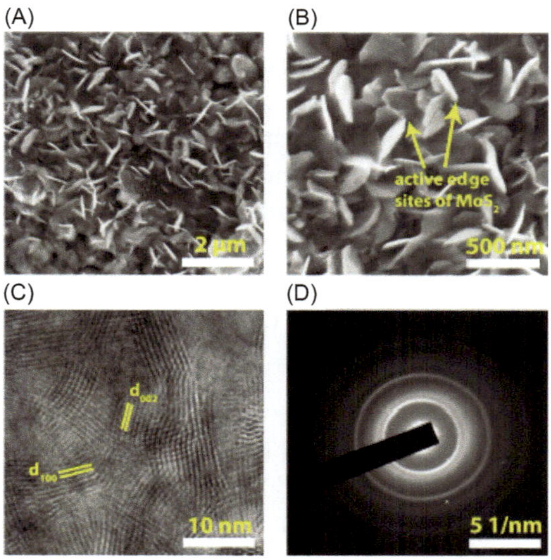

Figure 5.3 (A) A SEM, (B) a high resolution SEM, (C) a TEM, and (D) the SAED pattern of MoS_2 synthesized at the optimal conditions. Reprinted from Ref. [33], Copyright 2016, with permission from Elsevier.

The MoS_2 synthesized at optimal condition achieves J_{sc} of 15.2 mA/cm^2, V_{oc} of 0.707 V, FF of 0.697, and conversion efficiency η of 7.50% [33]. These photovoltaic parameters are higher than DSSCs using Pt CEs (14.6 mA/cm^2, 0.712 V, 0.7, and 7.28%, respectively).

The MoS_2 films prepared at different precursor concentrations show the marigold flower-like nanostructures of dimensions 100–300 nm, reported by R. Senthilkumar et al. [39]. At lower precursor concentrations, infolded nanostructures resembling petals are observed, while at higher precursor concentrations, these petal-like lamellae are bunched up to form marigold flower-like nanostructures due to crystal growth and ripening [39]. This indicates that the precursor concentration plays a significant role in controlling the morphology of MoS_2 thin films, which in turn affects the catalytic property of MoS_2 [40, 41]. The MoS_2 films prepared at different precursor concentrations [MoS_2-x (x = A: 3.2, B: 6.4, C: 9.6, and D: 12.6 mM)] show the marigold flower-like nanostructures of dimensions 100–300 nm. The best performing counter electrode is MoS_2-C, and it shows a power conversion efficiency of 6.32%, an open-circuit voltage (V_{oc}) of 0.75 V, J_{sc} of 15.55 mA/cm^2, and FF of 0.54, and under one sun illumination. The performance of this electrode is only slightly lower than that for Pt CE, which shows a conversion efficiency of 6.38% under the same condition. The performance of DSSC CEs is also assessed using incident photon-to-charge carrier efficiency (IPCE) spectra in the range of 400–750 nm. The CE curves of the DSSCs assembled with MoS_2-A, MoS_2-B, MoS_2-C, MoS_2-D, and Pt counter electrodes (as reference) show the IPCE values of 16.60%, 35.20%, 65.84%, 64.72%, and 68.38%. The best performing cell is MoS_2-C, with performance remarkably close to the Pt-based cell. These cells show the peak photoelectric response at 520 nm.

5.3 Ternary Transition Metal Sulfides

Till now, only Co/Ni-based ternary sulfides have been explored as CE materials in DSSCs [42–45]. Recently, Ag_8SnS_6 CE is fabricated for DSSCs and shows an impressive electrocatalytic activity and stability in catalyzing I_3^- reduction [46]. However, the photovoltaic performance of DSSCs based on Ag_8SnS_6 CE is still poor compared with that of Pt CE. Cartledge et al. and other researchers considered that the higher ionic potential could cause higher polarizability [46, 47–49], which could promote the adsorption of I_3^- on CE materials and enhance the electrocatalytic efficiency of CE in DSSCs. So

Ag_8GeS_6 would be a potential CE material due to its higher ionic potential (effective static potential of the cation) of Ge^{4+} compared to Sn^{4+}. Recently, Zhang et al. reported the colloidal process to synthesize Ag_8GeS_6 nanocrystals using GeI_4 as the Ge precursor [50]. Here, Ag_8GeS_6 nanocrystal ink is prepared in a mixture of oleylamine (OAm) and 1-octadecene (ODE) with $AgNO_3$, $GeCl_4$, and sulfur as precursors to fabricate CE for DSSCs for the first time. It is noteworthy that the DSSC with Ag_8GeS_6 CE displayed a power conversion efficiency of 8.10%, which was superior to that with Pt CE (8.02%). Two pairs of redox peaks observed in CV correspond to OX-1/Red-1 (left pair) and the Ox-2/Red-2 (right pair), which correspond to the redox reactions of I^-/I_3^- ($I_3^- + 2e^- \rightarrow 3I^-$) and I_2^-/I_3^- ($3I_2 + 2e- \rightarrow 2I_3^-$), respectively [51, 52]. The cathodic peak current density (J_{Red}) and peak-to-peak separation (E_{pp}) between anodic and cathodic peaks are important parameters for the electrocatalytic activity of CEs [51, 29, 53]. The higher J_{Red} (1.81 mA/cm^2) and lower E_{pp} (0.31 V) suggest that Ag_8GeS_6 CE possesses better electrocatalytic activity than that of Pt CE (1.75 mA/cm^2 and 0.35 V, respectively). From the EIS measurements, the smaller R_{ct} of Ag_8GeS_6 CE indicates its higher electrocatalytic activity for the reduction of I_3^- in DSSCs, which could promote the dye regeneration on photoanode and further improve the photocurrent density of DSSCs. The greater slope of the *J–V* plot in Tafel curves indicates higher J_0 of Ag_8GeS_6 CE than that of Pt CE. However, the corresponding constant phase angle element (CPE) value for the Ag_8GeS_6 CE is 10.02 μF, which is lower than that of the Pt CE with a CPE value of 10.41 μF due to the smaller particles of Pt, resulting in the larger surface roughness of Pt CE [54]. Taking the results of CV, EIS, and Tafel polarization into consideration, Ag_8GeS_6 CEs displayed higher electrocatalytic activity towards the reduction of I_3^- compared with Pt CEs, making them potentially good performer for constructing DSSCs. The photovoltaic performances demonstrate that Ag_8GeS_6 CE possesses better performance in DSSCs than Pt. This can be further supported by the IPCE spectra. The *J–V* characteristics show that DSSCs with the Ag_8GeS_6 CE retained 81.8%, 95.4%, and 78.1% of their original values, while ones with Pt CE are of 69.7%, 92.4%, and 65.4% (5, 10, 15 days), respectively. If the fabricated CEs were dipped in the iodine-based electrolyte for 15 days and then assembled into DSSCs, the retention rate of cells with Ag_8GeS_6 CE (95.6%) is higher than that of Pt CE (91.4%).

In the case of MIn_2S_4 (M = Fe, Co, Ni), the CV shows a pair of redox peaks which refers to the reaction of $I_3^- + 2e^- \leftrightarrow 3I^-$ on the surface of CEs [55, 56]. The higher reduction peak current density (J_{red}) and lower E_{pp} are the intuitive reflections of the excellent catalytic activity and electron transfer efficiency for CEs [57, 58]. The J_{red} is in the order of $CoIn_2S_4 > NiIn_2S_4 > FeIn_2S_4$ and the E_{pp} values are in the reverse order, thus revealing that $CoIn_2S_4$ CE possesses the highest catalytic activity for I_3^- reduction. The excellent catalytic activity of $CoIn_2S_4$ CE can be mainly attributed to its largest specific surface area (144.07 m^2/g), which can provide the most catalytic activity sites, and its abundant hierarchical pore structure, which leads to the highest charge transfer rate. The reversibility of the redox reaction ($I_3^- + 2e^- \leftrightarrow 3I^-$) can be evaluated by the ratio of $|J_{ox}/J_{red}|$ and the ratio close to 1 indicates the excellent reversibility. The J_{ox} is the oxidation peak current density. $|J_{ox}/J_{red}|$ ratios for different CEs are calculated and in an order of $CoIn_2S_4$ (1.01) < $NiIn_2S_4$ (1.20) < $FeIn_2S_4$ (1.22), indicating a reversible order of $CoIn_2S_4 > NiIn_2S_4 > FeIn_2S_4$ and giving the conversion rate (from I_3^- to I^-) order of $CoIn_2S_4 > NiIn_2S_4 > FeIn_2S_4$. The catalytic activity of binary references complies with the order of $CoS_2 > NiS/Ni_3S_4 > FeS_2 > In_2S_4$. From the CV measurements of different CEs at different scan rates, one can conclude that J_{red} is the linear function of the square root of scan rate, indicating that the I_3^- reduction is mainly controlled by the ionic diffusion process in the electrolyte [59, 60]. From EIS, the attenuation trend of R_{ct} is $FeIn_2S_4 > NiIn_2S_4 > CoIn_2S_4$, which indicates an inverse order of charge transfer rate and catalytic activity of CEs. Tafel plots show $CoIn_2S_4$ CE having the best catalytic activity due to its highest limiting diffusion current density (J_{lim}) and the J_0. DSSC based on the $CoIn_2S_4$ CE achieves the better PCE of 8.83%, with J_{sc} of 17.65 mA/cm^2, V_{oc} of 0.77 V, and FF of 0.65, compared to $FeIn_2S_4$ (7.18%) and $NiIn_2S_4$ (8.27%) CEs. The largest J_{sc} of $CoIn_2S_4$ can be attributed to its most catalytic activity sites and specific nanosheet array channel structure. Compared with $CoIn_2S_4$ and $NiIn_2S_4$ CEs, the lowest FF of $FeIn_2S_4$ CE has resulted due to its lowest surface area and the worst nanosheet array structure. Moreover, the PCEs based on ternary MIn_2S_4 CEs (M = Fe, Co, Ni) are comparable with that of Pt CE (8.19%), which are higher than those of their corresponding binary references: In_2S_4 (0.46%), FeS_2 (6.37%), CoS_2 (7.15%), and NiS/Ni_3S_4 (6.62%), respectively.

Furthermore, the influence of the thickness of $CoIn_2S_4$ on the catalytic and the device performances is investigated via optimizing the solvothermal reaction time. According to the highest $|J_{red}|$ and the lowest E_{PP}, $CoIn_2S_4$ CE with the thickness of 950 nm achieved the highest catalytic activity and the best photovoltaic performances (8.83%) with a solvothermal reaction time of 12 h.

The CV curves of NG, Ni-Mo-S/NG hybrid, and Pt [61] show a pair of strong redox peaks corresponding to the oxidation and reduction reactions and are given by the following equations:

$$3I^- \rightarrow I_3^- + 2e^- \text{ (Oxidation)} \tag{5.1}$$

$$I_3^- + 2e^- \rightarrow 3I^- \text{ (Reduction)} \tag{5.2}$$

It should be noted that the separation between anodic and cathodic peaks (E_{pp}) for Ni-Mo-S/NG is (~212 mV) significantly smaller than for the other CEs (NG = ~320 mV, pure Ni-Mo-S = ~252 mV, and Pt = ~301 mV). The higher current density and the lower E_{pp} properties of Ni-Mo-S/NG are directly related to the enhancement of electrocatalytic activity towards I^-/I_3^- redox reaction. The increase in the efficiency of DSSCs is a result of the hierarchical architecture and the higher number of wide-open edge sites on the NG support. Upon Ni-Mo-S anchoring onto the NG framework, Ni-Mo-S/NG exhibits excellent performance, showing the highest peak current density and the lowest E_{pp} when compared to other CEs. Further enhancement of electrocatalytic activity requires the formation of Ni-Mo-S with a higher number of electroactive sites and greater extend of doping of nitrogen into the graphene framework during the synthesis process.

The CV curves at a different scan rate for Ni-Mo-S/NG with a voltage window from −0.4 to 0. 4 V show strong redox peaks as well as increased current densities with respect to the scan rates from 5 to 100 mV/s. When the scan rate increases from 5 to 100 mV/s, there is no change observed in the shapes of these CV curves, indicating improved electron transport kinetics, remarkable electronic conductivity, superior rate performance, and low R_{ct}. The linear relationship between the square root of scanning rates and the peak (cathodic and anodic) current densities indicates that the oxidation and reduction reactions are diffusion limited for the transport of I^-/I_3^- species in the direction of the CE in the DSSC device. The CV curves of NG and Ni-Mo-S/NG did not show significant changes in anodic and cathodic peak potentials. In contrast, the Pt CE shows significant changes after 100 consecutive cycles in cathodic and anodic peak current densities. These results confirm that Ni-Mo-S/NG CE possesses exceptional reversible redox properties, outstanding electrochemical stability, and robust adhesiveness on the FTO conducting substrate.

EIS of all CEs show two distinctive semicircles. The NG CE shows a lower R_{ct} value (0.96 Ωcm^2) compared to Pt CE (2.18 Ωcm^2). The lower R_{ct} of NG CE is a result of exceptional electronic conductivity and many electrocatalytic Ni-Mo-S active sites. The anchoring of 2D Ni-Mo-S with nano-architectures into NG matrices leads to a significantly reduced R_{ct} (0.07 for Ni-Mo-S/NG), representing its superior electrocatalytic activity as redox shuttle in DSSCs. From the Tafel polarization study, the Ni-Mo-S/NG CE is found to have the highest J_0, confirming its high efficiency for the reduction process of I_3^- to I^- in the redox shuttle in DSSCs compared to Pt CE. This Tafel polarization result is well consistent with the CV and EIS studies. The performed electrochemical studies, including CV, EIS, and Tafel polarization, show that Ni-Mo-S/NG CE has superior electrocatalytic activity towards I_3^- reduction compared to Pt CE. This is a result of highly synergistic interactions between Ni-Mo-S and NG nanosheets, which result from its hierarchical hybrid architecture and unique porous network. The J–V characteristic shows that the power conversion efficiency of hierarchical Ni-Mo-S/NG hybrid (~9.89%) is noticeably higher than that of the recently reported transition metal chalcogenide (TMC)-based CE in DSSCs. Such exceptional photovoltaic performance of Ni-Mo-S/NG hybrid is anticipated from its hierarchical 2D Ni-Mo-S nanosheets with unique structures in the NG framework. The ultra-thin Ni-Mo-S nanosheets are well known to have a perfect adsorption energy for the I^- medium, leading to naturally superb electrocatalytic activity and electrochemical stability towards the reduction of I_3^- in DSSCs [62].

Tafel polarization analysis of Cu_2SnS_3 (CTS) shows a larger slope, suggesting a higher J_0 value and thus a smaller R_{ct} on the CTS electrode [63]. It is observed that the J_{lim} value of CTS electrode is slightly higher than that of Pt and hence inferring the increase in the catalytic activity of the electrode [27]. From CV studies, it is found that the E_{pp} of the CTS electrode (260 mV) is slightly smaller than that of Pt (340 mV) and is directly responsible for the fact that the overpotential loss of the CTS CE is lower than that of the Pt CE [64, 65]. After 10 cycles of scan, the R_{ct} value of the CTS CE approached that of fresh Pt (0.75 μcm^2) and still maintained about 80% of its initial value after 19 cycles. The increase in the R_{ct} value, that is, the deactivation of CE in I_3^-/I^- medium when subjected to repeated scanning of EIS was also observed in the case of the standard Pt, except that the latter exhibits a slower decline rate. This deactivation is interpreted as being due to the poisoning effect of the catalyst.

The CV curves of different counter electrodes of $CoNi_2S_4$ (NCS-20 min, NCS-90 min, NCS-300 min) under I_2^-/I_3^- electrochemical system using a three-electrode method exhibit two pairs of redox peaks. The CV curve of NCS-20 min is quite different from the other curves and there are no redox peaks at more negative potentials, indicating the poor electrocatalytic properties of the sample toward I_3^-/I_2^- reduction [66, 67]. The other two samples NCS-(90 min) and NCS-(300 min) show electrocatalytic ability for reducing triiodide ions. To understand the charge transfer process at the interface between the electrolyte and the counter electrode, we measured the electrochemical impedance spectra (EIS) by using the symmetric sand-which type configuration in order to eliminate the influence of the photoanode. Among all the counter electrodes, NCS-20 counter electrode has the largest semicircle in the Nyquist plot, suggesting that there is a large R_{ct}, which is in accordance with the poor performance of reducing I_3^-/I_2^- ions in the CV measurement. The R_{ct} value of the NCS-90 min is lower than that of the NCS-300 min. The Pt counter electrode has the lowest R_{ct} value. The NCS-90 has the lowest sheet resistance and the other counter electrodes have similar values. This demonstrates that $CoNi_2S_4$ has a prominent conductivity. Due to the increased thickness of the deposited film, the sheet resistance of the NCS-300 is larger than that of the NCS-90. Based on the above results, the NCS-90 has the best catalytic capability and the lowest R_{ct} due to its proper film thickness. If the thickness is very small (NCS-20), the low amount of nanoparticles cannot provide enough catalysis sites for the chemical reaction, and if the thickness is very large (NCS-300), the I_2^-/I_3^- solution cannot penetrate through the pores located in the films, and thus the catalytic effect is also decreased. DSSCs based on NSC-20 counter electrode shows a low FF as a result of its poor electrocatalytic property. As for the NSC-90 counter electrode, J_{sc} and FF values of DSSCs increase obviously, leading to enhanced conversion efficiency from 2.49% to 4.61%. Both electrodes show excellent electrocatalytic property, and the low sheet resistance contributes to the enhanced conversion efficiency. However, the efficiency of NSC-300 CE-based DSSCs decreases significantly due to the decreased J_{SC}. This can be attributed to the increased charge transfer resistance at the counter electrode as confirmed by the EIS analysis. In short, the NSC-90 counter electrode exhibits the best performance among the three different NSC nanostructures and its conversion efficiency is comparable to that of the Pt CE-based DSSCs. In spite of a less satisfactory electrocatalytic property of the Pt counter electrode, DSSCs using Pt as the catalyst still possess the highest conversion efficiency among all the DSSCs. It should be pointed out that the conversion efficiency of DSSCs is less than that of typical DSSCs due to the thickness of TiO_2 photoanode (8 mm) and lack of the scattering layer.

5.4 Quaternary Transition Metal Sulfides

Cu_2ZnSnS_4 (CZTS): Among the inorganic compounds, quaternary chalcogenide semiconductor compounds have drawn greater attention in the field of photovoltaics due to the abundance of their constituent elements. Recent studies reported the use of Cu_2ZnSnS_4 (CZTS) and $Cu_2ZnSnSe_4$ (CZTSe) nanocrystalline films in DSSC and quantum dot dye sensitized solar cells (QDSSCs) as counter electrodes [68, 69]. DSSCs with cost-effective quaternary sulfide counter electrode were prepared by K. Mokurala et al. [68]. Figure 5.4 shows FEG–TEM images of CZTS nanoparticles synthesized at 200°C for different reaction times (24 h, 36 h) for different solvents. For the solvent as ethylene glycol (EG), a mixed morphology consisting of nanorods with nanosheets is observed for the reaction carried out at 200°C for 24 h, while nanoflower with nanosheets is observed for a prolonged reaction of 36 h and deionized water (DW) as the solvent. The mixed morphology of nanorods with sheets might be possibly due to the adsorption of hydroxyl groups of EG over the surface of CZTS particles. This retards the growth of nanoparticles, thus allowing the growth only in a specific direction. Electrocatalytic activity of Pt- and CZTS- (S: spherical nanoparticles, R: nanorods, F: flowers) based CEs is evaluated using cyclic voltammetry. Cyclic voltammetry is carried out using a three-electrode system in an iodide-based (I_3^-/I^-) electrolyte solution. Cyclic voltammetry (CV) of these CEs show pairs of two redox peaks as shown in Fig. 5.5. The first pair (ox1 and red1) is attributed to the redox reaction of I_3^-/I^-, while the second pair (ox2 and red2) corresponds to the redox reaction of I_2/I_3^-. However, the more negative (first) pair of peaks (ox1 and red1) which is responsible for the regeneration of the redox couple (I_3^-/I^-) in DSSCs is considered. Cathodic current density (I_c) and E_{pp} are the two important parameters for the evaluation of the catalytic activity of different CEs.

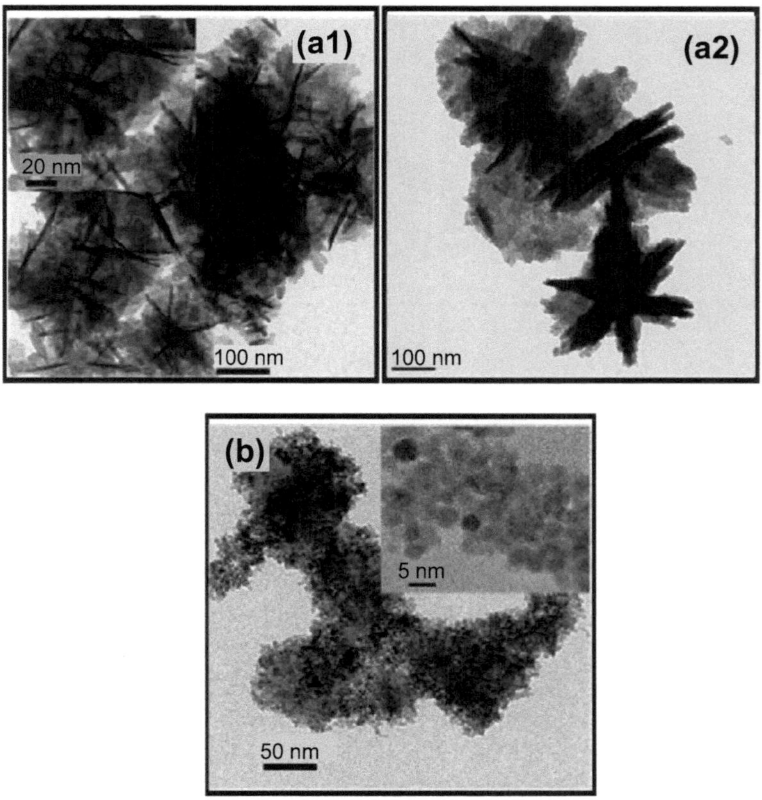

Figure 5.4 FEM-TEM images of CZTS nanoparticles synthesized at 200°C for (a1) 24 h and (a2) 36 h using EG as the solvent. (b) Morphology of CSTS nanoparticles remains the same for prolonged reaction time [24–36 h for deionized water (DW)]. Reprinted from Ref. [68], Copyright 2016, with permission from Elsevier.

CZTS- (R, F, and S) based CEs show pairs of two redox peaks similar to that of Pt CE (Fig. 5.5). The Pt-based CE has the highest cathodic peak current density (I_c = 1.48 mA/cm^2) and the lowest E_{pp} (325 mV) among all the other CEs. The CZTS-R-based CE is observed to have the highest I_c (1.42 mA/cm^2) and the lowest E_{pp} value (493 mV) among all the CZTS-based CEs. Similar CVs and electrocatalytic parameters to that of Pt make CZTS a promising CE material capable of effective reduction of I_3^- to I^- at the CE and electrolyte interface.

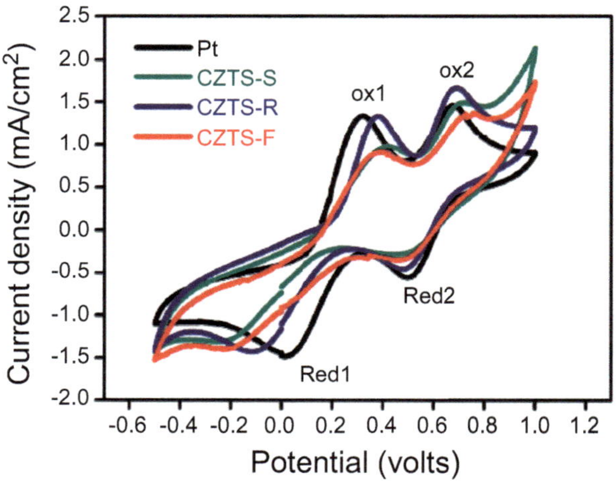

Figure 5.5 Cyclic voltammograms of CZTS- (S: spherical nanoparticles, R: nanorods, F: flowers) and Pt-based CEs. Reprinted from Ref. [68], Copyright 2016, with permission from Elsevier.

The J–V analysis shows that DSSCs fabricated with CZTS-R-based CE exhibited the largest short-current density (J_{sc} = 18.20 mA/cm^2 and η = 7.8%) among all the CZTS-based CEs. This observed J_{sc} for CZTS-R-based CE is comparable with the Pt-based CE (18.50 mA/cm^2) [68, 70].

Among the inorganic materials, quaternary chalcopyrite sulfides/selenides such as CZTS (Cu_2ZnSnS_4), CZTSe ($Cu_2ZnSnSe_4$), and CFTS (Cu_2FeSnS_4) have already been explored as CE materials [69–73]. Chalcopyrite sulfides are compounds composed of nontoxic, earth-abundant elements [74]. It has the stannite crystal structure with space group I-42m [74]. It is noted that resistivity values obtained from the Hall measurements and four-probe techniques are in good agreement with Pt CE. Higher resistivity and lower carrier mobility of CFTS film as compared to CCTS film might be due to elementary defects (Cu or S) in CFTS films. These defects can trap the electrons and increase the resistivity of CFTS film. Similar kind of observation has been reported for CZTS film [75–77]. The cyclic voltammetry of Pt, CCTS, and CFTS film is carried out using redox electrolyte solution (LiI/I_2) in a three-electrode system set-up. Two pairs of redox peaks are observed in the cyclic voltammograms of the prepared CCTS, CFTS films, which are similar to that of Pt film. The pair with negative potential is attributed to the redox reaction of I_3^-/I^-, which is responsible for the regeneration of the dye in DSSCs. However, the pair with positive potential corresponds to the redox reaction of I_2^-/I_3^-. Peak current density (I_{pc}) and E_{pp} are the two important parameters to evaluate the electrocatalytic activity of different CEs. The rate conversion of I_3^-/I^- is inversely proportional to the E_{pp} value [29, 57]. The electrocatalytic parameters, such as peak current density (I_{pc1}, I_{pa1}), E_{pp}, and J, as calculated from the CV curves of different CEs show that CCTS has a better electrocatalytic activity to reduce I_3^- ions in the electrolyte solution as compared to CFTS, but smaller than that of Pt CE. According to the existing literature, specific surface area, morphology, and elemental composition of quaternary chalcopyrite sulfides (CZTS and CZTSe) affect the electrocatalytic parameters [75, 76, 78]. J–V curves of the DSSCs fabricated with CCTS, CFTS, and Pt-based CEs show lower short-circuit current density of CFTS-based DSSCs, which can be attributed to the higher resistivity and lower specific surface area of CFTS CE [75, 78]. The efficiencies of the CCTS, CFTS, and Pt CE are found to be 7.4%, 7.1%, and 8.2%, respectively. The IPCE of DSSC with CCTS CE is observed to be about 83% at 520 nm, which is close to that of Pt (87% at 520 nm), in contrast to CFTS CE (78% at 520 nm). The IPCE spectra results are in accordance with the photoconversion efficiencies of DSSCs fabricated using different CEs [71].

Also, Cu_2FeSnS_4 (CFTS), Cu_2-$CoSnS_4$ (CCoTS), Cu_2CdSnS_4 (CCdTS), $Cu_2ZnSnSe_4$ (CZTSe), and Cu_2ZnSnS_4 (CZTS) have shown similar electrochemical catalytic activity to Pt and have been explored as a CE material in DSSCs [72, 72, 79–82]. The electrical properties of all CCdTS and CFTS films were evaluated by using Hall measurements and found to show P-type conductivity. The resistivities of sulfurized CFTS and CCdTS films are smaller than those of the films annealed in N_2 atmosphere. The mobility of sulfurized films is higher than that of the films annealed in N_2 atmosphere. The higher resistivity and the lower mobility of CCdTS film are due to changes in morphology and deviation in the stoichiometric ratio of CFTS films. The deficiency of S and Sn creates elementary defects V_s (sulfur vacancy) and V_{Sn} (tin vacancy) in the prepared films, which can trap electrons, thus increasing the resistivity of the films [75, 83]. The obtained electrical properties of CCdTS and CFTS films are found to be similar to the reported values (Table 5.1) [84].

Table 5.1 Electrical properties of CCdTS and CFTS films

Parameters	N-CFTS	N-CCdTS	S-CFTS	S-CCdTS
Type of conductivity	P	P	P	P
Mobility [$cm^2/(V\,s)$]	1.7	1.2	4.5	3.2
Carrier concentration (cm^{-3})	3×10^{17}	2.5×10^{17}	3.3×10^{17}	2.7×10^{17}
Resistivity (Ω cm)	0.56	0.98	2×10^{-2}	7×10^{-2}

Cyclic voltammetry measurements of prepared films show two pairs of redox peaks, indicating that synthesized materials can act as the electrocatalyst for the reduction of I_3^- to I^-. The electrochemical catalytic parameters such as cathodic current density (I_{pc1}) and E_{pp} for different CEs are obtained from cyclic voltammograms. The higher I_{pc1} and lower E_{pp} values of sulfurized (S–CFCTS and S–CFTS) CE indicate a better electrocatalytic activity for the reduction of I_3^- ions in the electrolyte solution compared to nitrogen (N–CFTS and N–CCdTS) CE, but inferior activity to that of Pt CE [73, 77, 78]. Cyclic voltammetry is carried out for 100 consecutive cycles at 50 mV/s scan rate to confirm the stability of S–CCdTS CE and S–CFTS CE in the iodide-based (I_3^-/I^-) electrolyte system. The CV curves of S–CCdTS and S–CFTS CEs (100 cycles) are found to be analogous to that of Pt CE. The unchanged shape of the CV curves shows the electrochemical stability of the S–CCdTS CE and S–CFTS CE in the iodide-based electrolyte [85]. Electrocatalytic activity of CEs is further confirmed by EIS and Tafel polarization measurements performed under illumination at zero bias voltage. Electrochemical parameters such as J_0 (obtained from Tafel polarization), R_s, R_{ct}, and

Table 5.2 Electrochemical and solar cell parameters of DSSC fabricated with different CEs

CES	R_s (Ωcm^2)[a]	R_{ct} (Ωcm^2)[a]	CPE (μF)[a]	R_{ct} (Ωcm^2)[b]	J_o (mA/cm)[b]	J_{sc} (mA/cm)	V_{oc} (mV)	FF	(%)
Pt	10.9	2.1	42	2.4	4.8	18.27	713.6	0.60	8.15
S-CFTS	11.37	2.9	38	3.3	3.7	17.37	705.2	0.57	7.36
S-CCdTS	12.30	3.8	39	4.1	3.1	16.38	695.6	0.56	7.12
N-CFTS	13.2	5.0	34	5.4	2.3	14.59	684.6	0.54	5.78
N-CCdTS	12.4	6.9	32	7.1	1.8	13.29	675.0	0.53	5.30

[a]Calculated from EIS spectra
[b]Calculated from Tafel polarization

the corresponding CPE at the electrolyte/CE interface estimated from the EIS spectra show that the R_{ct} values of different CEs calculated from the EIS spectra and Tafel polarization curves are in good agreement with each other. The obtained R_{ct} and J_0 values indicate that sulfurized CEs are more efficient for the reduction of I_3^- to I^- at the electrolyte–CE interface as compared to the N_2 annealed CEs; however, it is less efficient in comparison to Pt CEs. The variation in R_s value for different CEs might be due to the change in their electrical resistivity and adhesion of the films to the FTO substrate [86–88]. The CV, EIS, and Tafel analysis indicate that the sulfurized CEs possess better electrochemical activity compared to the N_2 annealed CEs. Average solar cell parameters such as J_{sc}, V_{oc}, FF, and η estimated from J–V characteristics of DSSCs assembled with all CCdTS, CFTS, and Pt-based CEs are given in Table 5.2.

5.5 Penternary Transition Metal Sulfides

Besides quaternary chalcogenide, a fifth metal ion can be incorporated into quaternary chalcogenides to convert to penternary chalcogenides, such as $Cu_2ZnSn(SeS)_4$, (CZTSeS), and $Cu_2XSn(SeS)_4$ ($X = Fe^{2+}$, Co^{2+}, Ni^{2+}, Mn^{2+}, etc.), which provide additional optical, photochemical, thermoelectric, and magnetic functions [89–91]. The first penternary chalcogenide-based counter electrode for DSSCs was developed by Xin and coworkers [69]. The device has good catalytic performance for regeneration of iodide from triiodide in a redox electrolyte and has an efficiency of 7.37%, which is even higher than that of Pt-based DSSCs (7.04%). Although this device is inefficient because it has lower electron transport efficiency, it is a proof-of-concept that DSSCs can be created with penternary chalcogenide-based counter electrodes such as $Cu_2CoSn(SeS)_4$ (CCTSeS) NCs. Farukozel et al. synthesized penternary chalcogenides such as CCTSeS and CZTSeS nanocrystals by a hot-injection method [92]. Cyclic voltammetry experiments of CCTSeS-based electrode show a much larger current density for I_3^- reduction than both CZTSeS and Pt-based electrodes, which implies a faster redox reaction rate and a better electrocatalytic activity for I_2^-/I_3^- redox couple on the CCTSeS-based CEs. This is due to the increase in crystallinity, the large active surface area, and the increased catalytic active sites of CCTSeS-based counter electrodes [93–95]. However, when Zn^{2+} ions are replaced with Co^{2+} ions in the kieserite structure, the CZTSeS electrode shows lower anodic and cathodic peak current densities compared to CCTSeS and Pt-based electrodes. This confirms lower electrocatalytic activity. Although CZTSeS-based electrodes with better catalytic performance than Pt-based electrode have been reported by other researchers [69], the observed results are found to be opposite. This difference is mainly caused by the characteristics of CZTSeS films, such as the existence of organic ligands on surface, worse catalytic properties, much larger particle size, and lower conductivity of CZTSeS nanocrystal [96, 97]. They also have weaker adhesion to FTO than that of Pt particles [98]. The catalytic activity data implies that CCTSeS nanocrystals could be used as an efficient counter electrode in DSSCs and could replace Pt. EIS measurements show well-defined single semicircles over the high frequency range related to the impedance of charge transfer process occurring at the counter electrode and electrolyte interface. The CCTSeS electrode has the lowest R_{ct} of 2.23 Ω, which is lower than that of Pt (14.75 Ω) and CZTSeS (128.34 Ω) electrodes. This result suggests higher electrocatalytic activity and lower recombination at the interfaces in DSSCs with CCTSeS NCs as CEs [68]. In contrast, the CZTSeS electrode exhibited huge R_{ct} values, indicating poor catalytic activity, which can be attributed to the organic ligand capped on the surface of NCs [96, 97]. Besides, the CCTSeS electrode has the lowest sheet resistance and other counter electrodes have similar values. This demonstrates that CCTSeS NC has a prominent conductivity. However, R_{ct} and R_s are not the only deciding factors for counter electrode performance. The double layer capacitance (C_{dl}) can also provide information on the catalytic activity of the counter electrode. The C_{dl} of CCTSeS (87 mF) is higher than that of other CEs and corresponds to the higher surface area, which is a crucial factor for high electrocatalytic activities. In view of the excellent electrocatalytic activity and lower charge transfer resistance, it is expected that the DSSC based on the CCTSeS electrode can achieve improved performance. The DSSC based on CCTSeS as a CE with a maximum efficiency of 6.47%, which is greater than that of both CZTSeS (3.187%) and Pt CE (5.41%).

5.6 Conclusion

In DSSC, counter electrode plays an important role in the reduction of trioxide electrolyte. The basic requirement of the counter electrode to be used in DSSC is perfect stability and high catalytic activity. The conventional Pt electrode satisfies the above requirements, but it is expensive. So the research is focused on most inexpensive new materials for counter electrode in DSSC. It is noted that some of the binary, ternary, quaternary, and penternary metal sulfides materials show high catalytic activity.

Table 5.3 Comparison of *J–V* characteristics of binary, ternary, quaternary, and penternary sulfides for DSSC

CEs	J_{sc} (mA/cm^2)	V_{oc} (V)	FF	η (%)	Reference
Pt	16.34	0748	0.66	8.02	[66]
		Binary sulfides			
VS_2	13.65	0.726	063	6.24	[3]
CNTs/VS_2	15.05	0.753	0.681	7.72	[2]
CoS_2	14.62	0.71	0.64	6.78	[10]
SnS_2	15.66	0.759	0.53	6.30	[19]
NiS_2	12.16	0.714	0.64	5.56	[25]
MoS_2	15.55	0.75	0.54	7.50	[33]
		Ternary sulfides			
Ag_8GeS_6	16.59	0.746	0.65	8.10	[46]
$Coln_2S_2$	17.65	0.77	0.65	8.83	[55]
NiMoS	14.28	0.76	0.77	8.36	[55]
NiMoS/NG	15.65	0.79	0.80	9.89	[61]
Cu_2SnS_3	19.26	0.73	0.72	10.18	[63]
$CoNi_2S_4$	8.862	0.756	68.7	4.61	[61]
		Quaternary sulfides			
Cu_2ZnSnS_4	18.20	0.706	0.64	7.80	[68]
Cu_2FeSnS_4	17.37	0.705	0.60	7.36	[68]
Cu_2CdSnS_4	16.38	0.695	0.56	7.12	[69]
		Penternary sulfides			
$Cu_2ZnSn(SeS)_4$	3.23	0.65	0.45	3.18	[89]
$Cu_2CoSn(SeS)_4$	6.14	0.61	0.57	6.47	[89]

To show that the material is suitable as a counter electrode, CV, EIS studies, and the photochemical parameters are analyzed to find the efficiency of the cell. From the analysis, it is found that the catalytic property of some of the binary, ternary, quaternary, and penternary metal sulfides matches with that of the Pt electrode and the efficiency matches with the efficiency of Pt electrode (Table 5.3). From the results, it is concluded that some of the binary, ternary, quaternary, and penternary metal sulfides can be a suitable replacement for Pt in DSSC.

Acknowledgment

We would like to thank Ms. B Anupama and Ms. K M Aishwaraya, Five Year Integrated M.Sc. Physics, Amrita Vishwa Vidyapeetham for their support.

References

1. H. Sun, D. Qin, S. Huang, X. Guo, D. Li, Y. Luo, Q. Meng, *Energy Environ. Sci.*, 4 (2011) 2630.
2. K. M. Aiswarya, T. Raguram, K. S. Rajni, *Polyhedron*, 176 (2020) 114267.
3. G. Yue, X. Liu, Y. Mao, H. Zheng, W. Zhang, *Mater. Today Energy*, 4 (2017) 58–65.
4. Y. Xiao, G. Han, Y. Li, M. Li, Y. Chang, *J. Mater. Chem. A*, 2 (2014) 3452–3460.
5. G. Yue, J. Wu, Y. Xiao, J. Lin, M. Huang, Z. Lan, *J. Phys. Chem. C*, 116 (2012) 18057–18063.
6. G. Yue, J. Wu, Y. Xiao, M. Huang, J. Lin, Z. Lan, *Energy*, 54 (2013) 315–321.
7. S. Yuasa, T. Nagahama, A. Fukushima, Y. Suzuki, F. Ando, *Nat. Mater.* 3 (2014) 868–871.
8. C. E. Banks, T. J. Davies, G. G. Wildgoose, R. G. Compton, *Chem Commun.*, 18 (2005) 829–841.
9. R. Bajpai, S. Roy, P. Kumar, P. Bajpai, N. Kulshrestha, J. Rafiee, N. Koratkar, D. S. Misra, *ACS Appl. Mater. Interfaces*, 3 (2011) 3884–3889.
10. J. C. Tsai, M. H. Hon, I. C. Leu, *Chem. Asian J.*, 10 (2015) 1932–1939.
11. J. Yang, C. X. Bao, K. Zhu, T. Yu, F. M. Li, J. G. Liu, Z. S. Li, Z. G. Zou, *Chem. Commun.*, 50 (2014) 4824–4826.

12. C. H. Yoon, R. Vittal, J. Lee, W. S. Chae, K. J. Kim, *Electrochim. Acta*, 53 (2008) 2890–2896.
13. X. Chen, Y. Hou, B. Zhang, X. H. Yang, H. G. Yang, *Chem. Commun.*, 49 (2013) 5793–5795.
14. S. H. Hsu, C. T. Li, H. T. Chien, R. R. Salunkhe, N. Suzuki, Y. Yamauchi, K. C. Ho, K. C. W. Wu, *Sci. Rep.*, 4 (2014) 6983.
15. L. Y. Han, N. Koide, Y. Chiba, T. Mitate, *Appl. Phys. Lett.*, 2004, 84, 2433–2435.
16. J. Y. Lin, J. H. Liao, *J. Electochem. Soc.*, 159 (2012) D65–D71.
17. C. W. Kung, H. W. Chen, C. Y. Lin, K. C. Huang, R. Vittal, K. C. Ho, *ACS Nano* 6 (2012) 7016–7025.
18. J. Jin, X. Zhang, T. He, *J. Phys. Chem. C.*, 118 (2014) 24877–24883.
19. Y. Bai, X. Zong, H. Yu, Z. G. Chen, L. Wang, *Chem. Eur. J.*, 20 (2014) 1–8.
20. F. Fabregat-Santiago, J. Bisquert, E. Palomares, L. Otero, D. Kuang, S. M. Zakeeruddin, *J. Phys. Chem. C*, 111 (2007) 6550–6560.
21. Q. Wang, J. E. Moser, M. Grätzel, *J. Phys. Chem. B*, 109 (2005) 14945–14953.
22. Q. Wang, S. Ito, M. Grätzel, F. Fabregat-Santiago, I. Mora-Sero, J. Bisquert, T. Bessho, H. Imai, *J. Phys. Chem. B*, 110 (2006) 25210–25221.
23. M. Wu, X. Lin, A. Hagfeldt, T. Ma., *Chem. Commun.*, 47 (2011) 4535–4537.
24. X. Lin, M. Wu, Y. Wang, S. Hagfeldt, T. Ma, *Chem. Commun.*, 47 (2011) 11489–11491.
25. C. Zhang, L. Deng, P. Zhang, X. Ren, Y. Li, T. He, *Int. J. Electrochem. Sci.*, 12 (2017) 4610–4618.
26. Z. Li, F. Gong, G. Zhou, Z. S. Wang, *J. Phys. Chem. C*, 117 (2013) 6561–6566.
27. A. Hauch, A. Georg, *Electrochim. Acta*, 46 (2001) 3457–3466.
28. J. Zheng, W. Zhou, Y. Ma, W. Cao, C. Wang, L. Guo, *Chem. Commun.*, 51 (2015) 12863–12866.
29. F. Gong, H. Wang, X. Xu, G. Zhou, Z. S. Wang, *J. Am. Chem. Soc.*, 134 (2012) 10953–10958.
30. X. Chen, J. W. Guo, Y. Hou, Y. H. Li, S. Yang, L. R. Zheng, B. Zhang, X. H. Yang, H. G. Yang, *RSC Adv.*, 5 (2015) 8307–8310.
31. L. Kavan, P. Liska, S. M. Zakeeruddin, M. Gratzel, *ACS Appl. Mater. Interfaces*, 6 (2014) 22343–22350.

32. B. Zhang, D. Wang, Y. Hou, S. Yang, X. H. Yang, J. H. Zhong, J. Liu, H. F. Wang, P. Hu, H. J. Zhao, H. G. Yang, *Sci. Rep.*, 3 (2013) 1836.

33. S. I. Raj, Z. Xu, W. Yang, F. Yang, L. Hou, and Y. Li, *Electrochim. Acta.*, http://dx.doi.org/doi:10.1016/j.electacta.2016.07.059

34. T. Y. Wang, L. Liu, Z. W. Zhu, P. Papakonstantinou, J. B. Hu, H. Y. Liu, M. X. Li, *Energy Environ. Sci.*, 6 (2013) 625.

35. Z. L. Hu, K. Xia, J. Zhang, Z. Y. Hu, Y. J. Zhu, *Electrochim. Acta*, 170 (2015) 39.

36. S. Jiang, X. Yin, J. T. Zhang, X. Y. Zhu, J. Y. Li, M. He, *Nanoscale*, 7 (2015) 10459.

37. S. H. Ahn, A. Manthiram, *Adv. Energy Mater.*, 6 (2016) 1501854.

38. R. Cisneros, M. Beley, J. F. Fauvarque, F. Lapicque, *Electrochim. Acta*, 171 (2015) 49.

39. R. Senthilkumar, Ramakrishnan, B. Murali, P. C. Ramamurthy, D. Kumaresan, N. K. Kothurkar, *J. Solid State Electrochem.*, doi:10.1007/s10008-018-4043-7.

40. B. Zhang, D. Wang, Y. Hou, S. Yang, X. H. Yang, J. H. Zhong, J. Liu, H. F. Wang, P. Hu, H. J. Zhao and H. G. Yang, *Sci. Rep.*, 3 (2013) 1836.

41. N. A. Jadhav, P. K. Singh, H. Rhee, B. Bhattacharya, *Nanoscale Res. Lett.*, 9(1) (2014) 575.

42. Z. Shi, K. Deng, L. Li, *Sci. Rep.*, 5 (2015) 9317.

43. J. Y. Lin, S. W. Chou, *Electrochem. Commun.*, 37 (2013) 11–14.

44. A. Banerjee, K. K. Upadhyay, S. Bhatnagar, M. Tathavadekar, U. Bansode, S. Agarkar, S. B. Ogale, *RSC Adv.*, 4 (2014) 8289–8294.

45. X. Zheng, J. Guo, Y. Shi, F. Xiong, W.-H. Zhang, T. Ma, C. Li, *Chem. Commun.*, 49 (2013) 9645–9647.

46. Q. He, S. Hang, C. Wang, Q. Qiao, N. Liang, M. Xu, W. Chen, *Chem. Commun.*, 49 (2013) 9645–9647.

47. G. H. Cartledge, Studies on the periodic system I. The ionic potential as a periodic function, *J. Am. Chem. Soc.*, 50 (1928) 2855–2863.

48. P. A. Cox, *Instant Notes Inorganic Chemistry*, Science Press, 2000.

49. K. P. Jensen, W. L. Jorgensen, *J. Chem. Theory Comput.*, 2 (2006) 1499–1509.

50. B. Zhou, Y. Xing, S. Miao, M. Li, W.-H. Zhang, C. Li, *Chem. Eur. J.*, 20 (2014) 12426–12431.

51. Z. Yang, M. Liu, C. Zhang, W. W. Tjiu, T. Liu, H. Peng, *Angew. Chem. Int. Ed.*, 52 (2013) 3996–3999.

52. M. Wu, X. Lin, A. Hagfeldt, T. Ma, *Angew. Chem. Int. Ed.*, 50 (2011) 3250–3254.
53. X. Zheng, J. Deng, N. Wang, D. Deng, W.-H. Zhang, X. Bao, C. Li, *Angew. Chem. Int. Ed.*, 53 (2014) 7023–7027.
54. Y. C. Wang, D. Y. Wang, Y. T. Jiang, H. A. Chen, C. C. Chen, K. C. Ho, H. L. Chou, C. W. Chen, *Angew. Chem. Int. Ed.*, 52 (2013) 6694–6698.
55. W. Hou, Y. Xiao, G. Han, *Angew. Chem. Int. Ed.*, doi:10.1002/anie.201705399.
56. J. D. Roy-Mayhew, D. J. Bozym, C. Punckt, I. A. Aksay, *ACS Nano*, 4 (2010) 6203–6211.
57. W. J. Hou, Y. M. Xiao, G. Y. Han, D. Y. Fu, R. F. Wu, *J. Power Sources*, 322 (2016) 155–162.
58. H. C. Sun, Y. H. Luo, Y. D. Zhang, D. M. Li, Z. X. Yu, K. X. Li, Q. B. Meng, *J. Phys. Chem. C*, 114 (2010) 11673–11679.
59. X. J. Zheng, J. Deng, N. Wang, D. H. Deng, W. H. Zhang, X. H. Bao, C. Li, *Angew. Chem. Int. Ed.*, 126 (2014) 7143–7147.
60. C. Bao, F. X. Li, J. L. Wang, P. P. Sun, N. Huang, Y. H. Sun, L. Fang, L. Wang, X. H. Sun, *ACS Appl. Mater. Interfaces*, 8 (2016) 32788–32796.
61. J. Balamurugan, G. Peera, M. Guo, T. T. Nguyen, N. H. Kim, J. H. Lee, *J. Mater. Chem. A*, 5 (2017) 17896–17908.
62. X. Cui, J. Xiao, Y. Wu, P. Du, R. Si, H. Yang, H. Tian, J. Li, W. H. Zhang, D. Deng, X. Bao, *Angew. Chem. Int. Ed.*, 126 (2016) 6820–6824.
63. F. Liu, S. Hu, X. Ding, J. Wen, X. Pan, S. Chen, Md-K. Nazeerudin, S. Dai, *J. Mater. Chem. A.*, 4 (2016) 14865–14876.
64. X. J. Zheng, J. Deng, N. Wang, D. H. Deng, W. H. Zhang, X. H. Bao, C. Li, *Angew. Chem. Int. Ed.*, 53 (2014) 7023–7027.
65. H. Wang, K. Sun, F. Tao, D. J. Sracchiola, Y. H. Hu, *Angew. Chem. Int. Ed.*, 52 (2013) 9210–9214.
66. Z. Shi, K. Deng, L. Li, *Sci. Rep.* 5 (2015) 9317.
67. T. Zhang, H. Chen, C. Su, D. Kuang, *J. Mater. Chem. A*, 1 (2013) 1724–1730.
68. K. Mokurala, A. Kamble, C. Bathina, P. Bhargava, S. Mallick, *Mater. Today Proc.* 3 (2016) 1778–1784 .
69. X. Xin, M. He, W. Han, Z. Lin, *Angew. Chem. Int. Ed.*, 50 (2011) 11739–11742.
70. S. J. Yuan, Z. J. Zhou, Z. L. Hou, W. H. Zhou, S. X. Wu, *Chem. Eur. J.*, 19 (2013) 10107–10110.

71. K. Mokurala, S. Mallick, P. Bhargava, *J. Power Sources*, 305 (2016) 134–143.

72. J. Y. Park, J. H. Noh, T. N. Mandal, S. H. Im, Y. Jun, S. Seok, *RSC Adv.*, 3 (2013) 24918–24921.

73. R. R. Prabhakar, N. H. Loc, M. H. Kumar, P. P. Boix, S. Juan, R. A. John, S. K. Batabyal, L. H. Wong, *ACS Appl. Mater. Interfaces*, 6 (2014) 17661–17667.

74. X. Zhang, N. Bao, K. Ramasamy, Y. H. A. Wang, Y. Wang, B. Lin, A. Gupta, *Chem. Commun.* 48 (2012) 4956–49582.

75. M. S. Fan, J. H. Chen, C. T. Li, K. W. Cheng, K. C. Ho, *J. Mater. Chem. A*, 3 (2015) 562–569.

76. S. K. Swami, N. Chaturvedi, A. Kumar, N. Chander, V. Dutta, D. K. Kumar, A. Ivaturi, S. Senthilarasu, H. M. Upadhyaya, *Phys. Chem. Chem. Phys.*, 16 (2014) 23993–23999.

77. V. G. Rajeshmon, C. S. Kartha, K. P. Vijayakumar, C. Sanjeeviraja, T. Abe, Y. Kashiwaba, *Sol. Energy*, 85 (2011) 249e255.

78. Y. Xie, C. Zhang, F. Yue, Y. Zhang, Y. Shi, T. Ma, *RSC Adv.*, 3 (2013) 23264–23268.

79. S. J. Yuan, Z. J. Zhou, Z. L. Hou, W. H. Zhou, R. Y. Yao, Y. Zhao, S. X. Wu, *Chem. Eur. J.*, 9 (2013) 10107–10110.

80. K. Mokurala, A. Kamble, A. Gupta, P. Bhargava, S. Mallick, *Adv. Sci. Lett.*, 22 (2016) 1026–1028.

81. K. Mokurala, A. Kamble, C. Bathina, P. Bhargava, S. Mallick, *Mater. Today*, 3 (2016) 1778–1784.

82. K. Mokurala, P. Bhargava, S. Mallick, *IEEE 40th Photovoltaic Specialist Conference (PVSC)*, 1 (2014) 1534–1537.

83. M. Kumar, A. Dubey, N. Adhikari, S. Venkatesan, Q. Qiao, *Energy Environ. Sci.*, 8 (2015) 3134–3159.

84. C. Rincon, M. Quintero, E. Moreno, Ch. Power, E. Quintero, J. A. Henao, M. A. Macıas, G. E. Delgado, R. Tovar, M. Morocoima, *Solid State Commun.*, 151 (2011) 947–951.

85. S. Yun, P. D. Lund, A. Hinsch, *Energy Environ. Sci.*, 8 (2015) 3495–3514.

86. H. Yuan, Q. Jiao, S. Zhang, Y. Zhao, Q. Wu, H. Li, *J. Power Sources*, 325 (2016) 417–426.

87. W. Hou, Y. Vaio, G. Han, R. Wu, *J. Power Sources*, 322 (2016) 155–162.

88. J. Song, G. R. Li, C. Y. Wu, X. P. Gao, *J. Power Sources*, 266 (2014) 464–470.

89. M. Gusain, P. Rawat, R. Nagarajan, *RSC Adv.*, 5 (2015) 43202–43208.
90. Q. Guo, G. M. Ford, W. C. Yang, B. C. Walker, E. A. Stach, H. W. Hillhouse, Rakesh Agrawal, *J. Am. Chem. Soc.*, 132 (2010) 17384–17386.
91. J. Shen, D. Zhang, J. Li, X. Li, Z. Sun, S. Huang, *Nano-Micro Lett.*, 5 (2013) 281–288.
92. F. Ozel, A. Sarilamz, B. Istanbulla, A. Ayabum, M. Kus, S. Sonmezogila, *Sci. Rep.*, 6, 29207 (2016) 1–9.
93. Y. Z. Zhagn, H. H. Li, D. X. Kou, W. H. Zhou, S. X. Wu, *RSC Adv.*, 5 (2015) 11599–11603.
94. S. Jeon, C. Kim, J. Ko, S. Im, *J. Phys. Chem. C.*, 115 (2011) 22035–22039.
95. Z. Tang, J. Wu, M. Zheng, Q. Tang, Q. Liu, J. Lin, J. Wang, *RSC Adv.*, 2 (2012) 4062–4064.
96. Y. R. Yao, Z. J. Zhou, Z. L. Hou, X. Wang, W. H. Zhou, S. X. Wu, *ACS Appl. Mater. Interfaces*, 5 (2013) 3143–3148.
97. J. Guo, X. Wang, W. H. Zhou, Z. X. Chang, X. Wang, Z. J. Zhou, S. X. Wu, *RSC Adv.*, 3 (2013) 14731–14736.
98. L. Han, N. Koide, Y. Chiba, A. Islam, R. Komiya, N. Fuke, A. Fukui, R. Yamanaka, *Appl. Phys. Lett.*, 86 (2005) 213501.

Chapter 6

Metal Chalcogenides as Counter Electrode Materials

Arunachalam Arulraj,[a,b] U. Mehana Usmaniya,[c] M. Ramesh,[c] J. Anandha Raj,[d] and G. Senguttuvan[b]
[a]*Graphene and Advanced 2D Materials Research Group (GAMRG), School of Science & Technology, Sunway University, Selangor, Malaysia*
[b]*University College of Engineering, Bharathidasan Institute of Technology (BIT) Campus, Anna University, Tiruchirappalli, India*
[c]*Smart Sensors, CSIR-Central Electronics Engineering Research Institute, Pilani, India*
[d]*Department of Energy Science, Alagappa University, Karaikudi, India*
sengutuvan@yahoo.com

Photovoltaic cells offer a best alternative to fossil fuels. Several approaches have been analyzed in order to have solar cells that are capable to conquer the energy market all around the world. Dye-sensitized solar cells (DSSCs), a modern generation photovoltaic, have drawn much attention owing to their ease of fabrication and potential for generating multiple electrons. Generally, DSSC has three main components in its architecture and one of the important components is counter electrode (CE), which plays a crucial role in collecting electrons from an external circuit and catalyzing the redox reaction in electrolyte, thereby profoundly impacting the

Counter Electrode for Dye-Sensitized Solar Cells
Edited by Alagarsamy Pandikumar, Ramesh Mohan, and Kandasamy Jothivnekatachalam

ISBN 978-981-4877-38-1 (Hardcover), 978-1-003-11077-4 (eBook)
www.jennystanford.com

performance of devices. So far platinum (Pt) is often considered as a superior CE material due to its excellent catalytic activity, chemical stability, and electrical conductivity; however, it has some drawbacks, such as less abundance ratio, expensive, and poor long-term stability against the iodine-based electrolyte, which limit its large-scale industrial applications in DSSCs. Therefore, economically beneficial replacement of Pt is important for the DSSC technology. The development of new materials (such as metal oxides/sulphides/nitrides, carbon-based materials, and chalcogenides) using cost-effective and earth-abundant elements for replacing Pt CEs for DSSCs, as well as the novel design in solar cells such as p-type DSSCs or tandem structures, has become technologically desirable. Among these different materials, chalcogenides (i.e., MX), where M is the metal (Mo, W, Ta, Nb, etc.) and X is a chalcogen (S, Se, Te), were made a progress on counter electrodes for developing Pt-free low-cost DSSCs with high photoconversion efficiency (PCE). They also possess the unique physical property of flexibility along with other properties. This chapter summarizes the special features and performance, various synthesis approaches, fabrications, characterizations, mechanisms, and merits and demerits of chalcogenide-based materials and their role as CE in DSSCs.

6.1 Introduction

As third-generation solar cells, DSSCs have stood out as the potential candidate for the traditional silicon-based solar cells due to their simple assembly procedure, good plasticity, environmental friendliness, and ease of building combination [1–3]. The concept of DSSCs was hypothesized by mimicking the photosynthesis system of the nature. Earliest DSSCs can be dated back to the pioneering work of H. Gerischer and H. Tributsch in the late 1960s, wherein the organic dyes adsorbed on ZnO single crystal electrodes could produce photocurrents [4, 5]. H. Tributsch and M. Calvin further demonstrated that sensitized ZnO semiconductors play a key role in solar energy conversion [5, 6]. At that time, PCE was low. It was not until the use of the rutile TiO_2 electrode in these electrochemical cells that the turn of the tide in the improvement of cell performance occurred [7, 8]. B. O'Regan and M. Gratzel then made a big breakthrough in 1991, which involved obtaining a high PCE of 7.1–7.9% in I^-mediated DSSCs assembled with a mesoporous anatase TiO_2 photoanode film, Pt CE, and a trimetric ruthenium complex dye [9, 10]. This stipulates a new era of R&D in the renewable solar energy technology. A research on DSSCs has bloomed into a race to develop high efficiency with low fabrication cost. To realize long-term developments and practical application of DSSCs, it requires a robust increase of PCE and a significant decrease in the production cost. In DSSCs, dye (sensitizer) is the master in improving PCE, and developing low-cost counter electrode as the catalyst is one of the key paths to reduce the cost of fabrication. Traditionally, Pt has been proven to be an outstanding CE catalytic material for DSSCs due to its high catalytic performance and excellent conductivity. Unfortunately, Pt is expensive and non-sustainable for long-term applications [11, 12]. Therefore, researchers are facing a real challenge of developing low-cost and earth-abundant alternatives.

Fast start-up and multiple start capabilities are two crucial requirements for the engines driven by solar cells. With this conceptual framework, researchers developed multiple varieties of materials, including metal oxides, nitrides, carbides, graphene-based materials, carbon-based materials, chalcogenides, and so on. Among all the novel materials, chalcogenides added one more royal feather on the hat of DSSCs by their unique properties. After the success of graphene, an equal emphasis was given to the development of two-dimensional (2D) materials that can form the atomic level of sheets with an extraordinary property. Interestingly, 2D library grows every year and features more than 150 exotic materials, which include different chalcogenides, such as molybdenum sulphides (MoS_2), molybdenum selenides ($MoSe_2$), tungsten sulphides (WS_2), and tungsten selenides (WSe_2) [13–15].

6.2 Role of Counter Electrodes

Before looking into Pt-free CEs, the importance and hamstrung in usage of alternative CEs need to be discussed by means of understanding the electrocatalytic process of redox mediators. The main task of CEs is to act as a catalyst by reducing redox species, which are the mediators for regenerating the sensitizer (dye) upon electron injection. Much of the research in DSSCs has focused on enhancing the performance of devices by boosting the short-circuit current density (J_{sc}), open-circuit voltage (V_{oc}), and fill factor (FF). By improving the CE material, the FF of the cell is increased, which is mainly influenced by the series resistance (R_s) of the cell related to the slope of the tangent line to the current density (J)–voltage (V) curve at V_{oc}. The series resistance is a combination of the Warburg impedance and the Nernst diffusion of the I_3^- species in the electrolyte (Z_N), which is the impedance related to charge transfer and recombination at the semiconductor–dye–electrolyte interface (R_{ct1}), resistance at the fluorine-doped tin oxide (FTO) glass (R_s), and the charge transfer resistance at the CE and the electrolyte interface (R_{ct2}). Equivalent circuit used to represent interfaces in DSSCs, where $R_{ct(solid)}$ is the charge transfer resistance of the semiconductor–FTO or Pt–FTO interface, its corresponding capacitance will be $C_{(solid)}$ and $W_{(sol)}$ is the Warburg parameter describing the diffusion of the I_3^- electrolyte.

Amongst all these resistances, a prominent role is handled by R_{ct} to determine R_s. A lower value of R_{ct} signifies tremendous electron transfer from the CE to the electrolyte for the reduction of I_3^- ion to I^- ion at the catalytic interfaces of the CE. A superficial electron transfer between the CE and the electrolyte reduces R_s, thereby giving a higher FF, resulting in high conversion efficiency [16–18].

The catalytic activity of the CE can be explained in terms of the current density (J), which is calculated from the charge transfer resistance (R_{ct}):

$$R_{ct} = RT/nFJ$$

where R, T, n, and F are the gas constant, temperature, the number of electrons transferred in the elementary electrode reaction ($n = 2$), and Faraday's constant, respectively [19]. The reactions at the CE mainly depend on the redox species used to transfer electrons between the photoanode and the CE. Usually, I^-/I_3^- couple is used as the redox mediator in DSSCs. The I_3^- produced due to regeneration of the dye is reduced at the counter electrode by the following reaction:

$$I_3^- + 2e^- \Leftrightarrow 3I^-$$

6.3 Chalcogenides

Chalcogenides (i.e., S, Se, and Te) belonging to VI A group in the periodic table create very stable compounds with transition elements and form layered crystalline structures specifically with transition metals of IV–VII B groups [IV B (Ti, Zr, Hf), V B (V, Nb, Ta), VI B (Mo, W), and VII B (Tc, Re)]. The layered structures formed with transition metals resemble graphite, with a weak van der Waals interaction between the individual layers. Depending on the atomic arrangement, the structures of chalcogenides can be categorized into trigonal prismatic (hexagonal, H), octahedral (tetragonal, T), and its distorted phase (T′), as shown in Fig. 6.1a [20], which give rise to dramatic anisotropy and different material properties. In general, chalcogen–metal–chalcogen arrangement along z-direction is considered as the single layer and weak van der Waals interactions take place in each layer (chalcogen–chalcogen), enabling mechanical exfoliation from bulk towards mono-/few-layer flake. T-phase has a trigonal chalcogen layer on the top and 180 degrees rotated structure, and so it is called “trigonal antiprism” at the bottom in a single layer and results in hexagonal arrangement of chalcogen atoms in the top view. Metal atoms are distorted further (or dimerized in one direction), called T′-phase [21, 22], resulting in the modification of atomic displacement of chalcogen atoms along the z-direction. Moreover, chalcogenides exhibit a variable bandgap depending on the elemental combination, the number of layers, and the lack or presence of doping atom from 0 eV (semimetal) to 2 eV (semiconductor). As shown in Fig. 6.1b, chalcogenide materials reveal a wide range of bandgaps covering the entire region of visible and infrared regimes with the choice of materials [23]. Most of the chalcogenides exhibit direct bandgap in single layer, whereas they are indirect in bulk form. Tuning of bandgap defines the electronic, electrochemical, and optical applications and tuning of selective doping helps in enriching the catalytic activity of chalcogenide-based materials. The structure and properties such as charge density wave, superconductivity, and magnetism are shown in Fig. 6.1c [20]. The detail description of all these properties in each material is beyond the scope of this chapter. Such kinds of properties are of great importance in broad areas of applications [24–26].

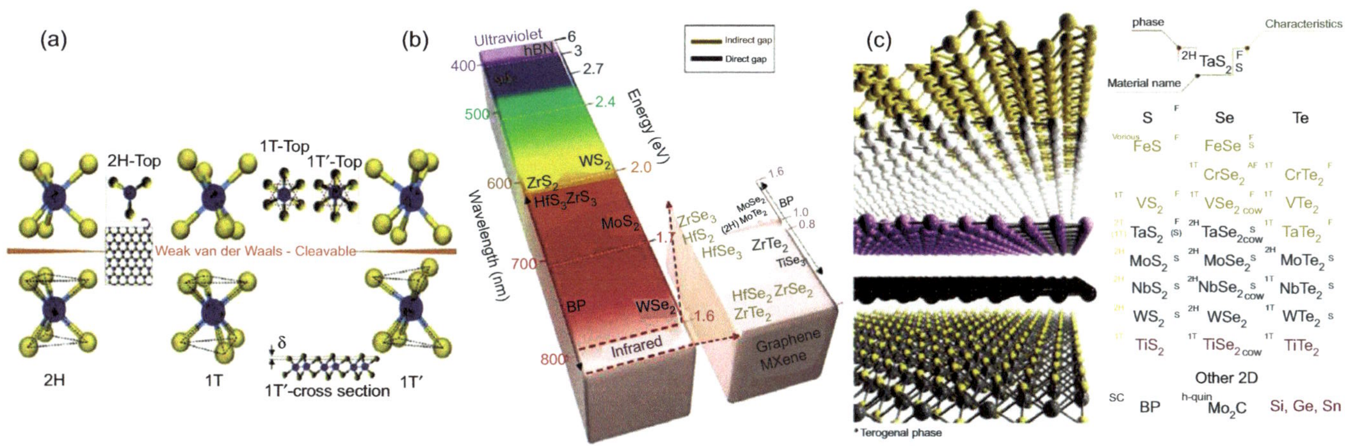

Figure 6.1 (a) Cleavable 2H, 1T, and 1T′ structures of layered transition metal chalcogenides, (b) bandgap of layered materials starting from graphene to hBN (hexagonal boron nitride) (zero to wide), and (c) various physical properties and the crystal structure of 2D TMDs and other 2D materials. Reprinted from Ref. [20], Copyright 2017, with permission from Elsevier.

In Fig. 6.1, the red colour denotes monolayer materials with stability under ambient conditions, blue indicates probability of stability at ambient atmosphere, and the green represents the unstable materials in ambient conditions, but stable at inert atmosphere. Among the various chalcogenide materials, MoS_2, $MoSe_2$, WS_2, TiS_2, $NbSe_2$, $TaSe_2$, $NiSe_2$, $FeSe_2$, $CoSe_2$, SnS_2, Bi_2Se_3, and so on are most often used in fabricating Pt-free CEs in DSSCs.

The chalcogenide-based CE acts an electrocatalyst, which is one of the significant components of DSSCs that facilitates the reduction of triiodide (I_3^-) to iodide (I^-) ions in redox electrolyte for dye regeneration. In this chapter, catalytic CEs developed from chalcogenide-based materials and their composites are discussed. The impact of material processing, fabrication, morphology, and electrochemistry of chalcogenide-based CEs for DSSCs are discussed in an elaborate manner.

6.4 Sulphide-Based Electrode

The pioneering work on using sulphides in energy applications started in the 1970s. In all extant, sulphide plays a crucial catalytic role in energy conversion systems. The potential connection between this essential role and the predominance of mineral sulphides has led to speculation in an emerging way. Another feature of sulphides that distinguishes them from oxides is their lower melting point in addition to being much more numerous than the corresponding oxides. Sulphide-based materials such as MoS_2 and WS_2 have attracted considerable attention as a CE material in the fabrication of DSSCs owing to their potential electrocatalytic activity [27]. The Young's modulus of monolayer MoS_2 outperforms the stainless steel (204 GPa) and graphene oxide (207 GPa), which is attributed to the absence of stacking faults, high crystallinity, and defect-free nature of the atomically thin TMDs [28].

MoS_2 and WS_2 chalcogenides found their application as electrocatalysts in DSSCs for the first time with PCE of 7.59% and 7.73%, respectively. The work was reported by M. Wu et al. in 2011. The authors synthesized both MoS_2 and WS_2 chalcogenides using the simple chemical synthesis approach and used as CE in DSSCs [29]. Researchers have paid more attention in developing these materials with different approaches, such as electrochemical, chemical bath deposition, hydrothermal, and pulsed laser deposition. V. H. V. Quy et al. fabricated MoS_2 via electrochemical deposition with changes in electrodeposition time (20, 40, 60 min). An optimized rate of deposition time (40 min) exhibited better electrocatalytic activity and showed much improved charge transport with PCE of 7.16% [30]. Similar to this work, S. Li et al. fabricated MoS_2/graphene on the flexible FTO substrate using electrochemical deposition as shown in Fig. 6.2a [31]. The authors varied the deposition time of graphene (20, 60, 80, 160 s) by fixing MoS_2 and achieved PCE of 8.01% for MoS_2/graphene (80 s) sample, which is comparable to that of the conventional Pt (8.31%). The authors also carried out the stability test by measuring the JV curve with periodic intervals of time. The JV curve and the optical photography of the MoS_2/graphene electrode are shown in Fig. 6.2b,c [31]. A. Arulraj et al. fabricated MoS_2 using pulsed laser deposition and achieved PCE of 5.663%, which is higher than that of the conventional Pt (~5%). The authors reported the effect of crystallinity and thickness in the PCE of the fabricated device. The author stated that with an optimized temperature conditions and varying the thickness of the films to particular extent, the PCE of the device can be increased drastically. The electrochemical and JV curves reported by the authors are given in Fig. 6.2d [32].

Tungsten disulphide (WS_2) belongs to the family of chalcogenide materials composed of the layered structure similar to that of MoS_2. An interesting fact of WS_2 is that the active sites are located on its edge rather than on its surface, which results in enhancing the electrocatalytic activity, thereby finding its application in the fabrication of DSSCs [33, 34]. Z. Shen et al. reported the synthesis of WO_x@WS_2@WO_x/carbon film by simple sulfurization process. The interconnected three-dimensional (3D) WS_2 structure with coated carbon films facilitates fast kinetic reactions in the electrolyte. WS_2 surface edge-oriented nanosheets facilitate electron transfer and electrolyte diffusion via carbon layer, and a PCE of 7.71% is achieved, which compares favorably with the conventional Pt (7.34%) [35]. Likewise, L. Keawphaisan et al. has made an attempt to synthesize WS_2–W_5O_{14}–MWCNTs using the hydrothermal approach. In this article, the author also highlighted the effect of annealing treatment of synthesized materials on the performance of the devices. Upon annealing of the fabricated devices at 450°C, there

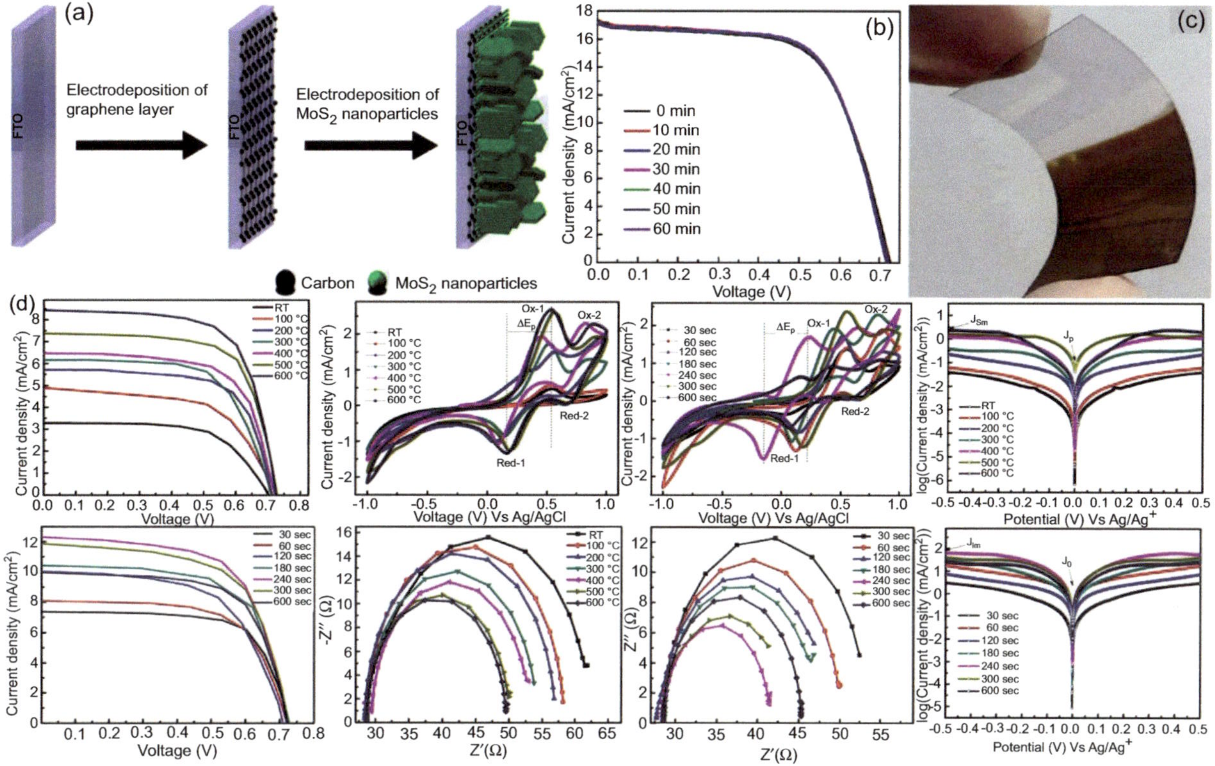

Figure 6.2 (a) Electrodeposition of MoS_2/graphene (a) Schematics; (b) JV curve and (c) Photograph of MoS_2 deposition on flexible PET substrate, reprinted from Ref. [31], Copyright 2016, with permission from the Royal Society of Chemistry; and (d) Photovoltaic and electrochemical characteristics of laser ablated MoS_2, reprinted from Ref. [32], Copyright 2018, with permission from Elsevier.

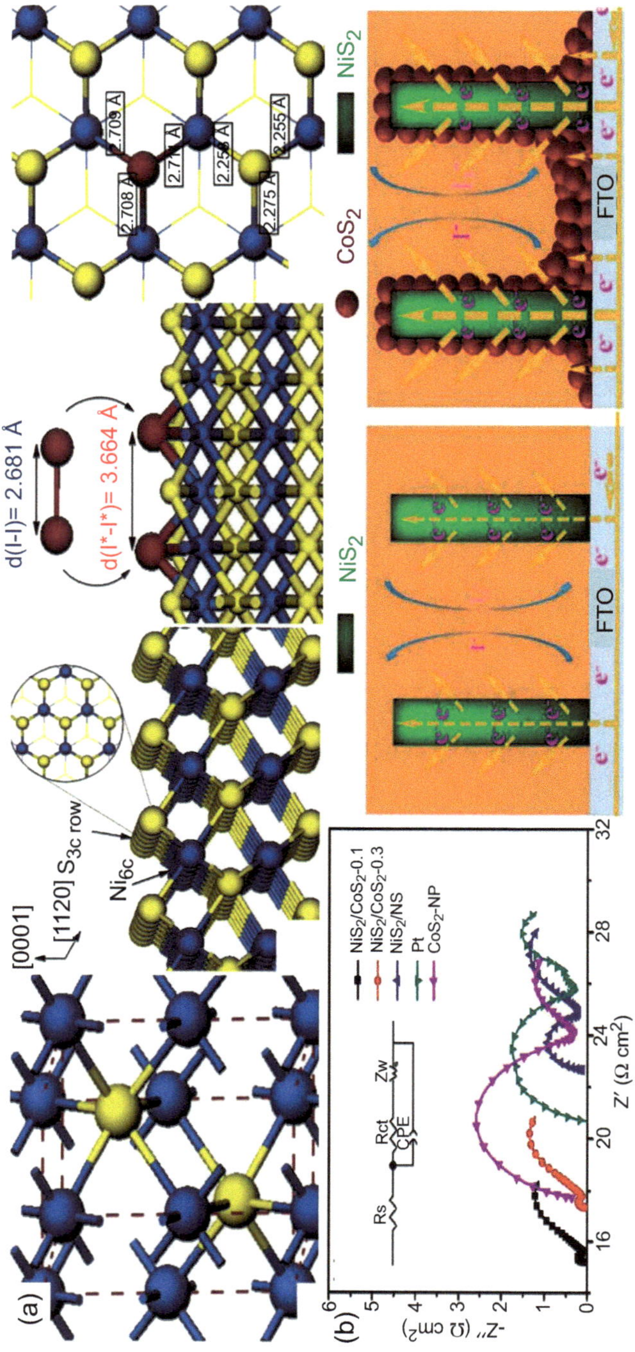
(a)
[0001]
[1120]
S3c row
Ni6c
d(I-I)= 2.681 Å
d(I*-I*)= 3.664 Å
2.709 Å
2.708 Å
2.256 Å
2.275 Å
2.255 Å
NiS2
CoS2
FTO
(b)
Rs
Rct
CPE
Zw
NiS2/CoS2-0.1
NiS2/CoS2-0.3
NiS2/NS
Pt
CoS2-NP
-Z'' (Ω cm2)
Z' (Ω cm2)

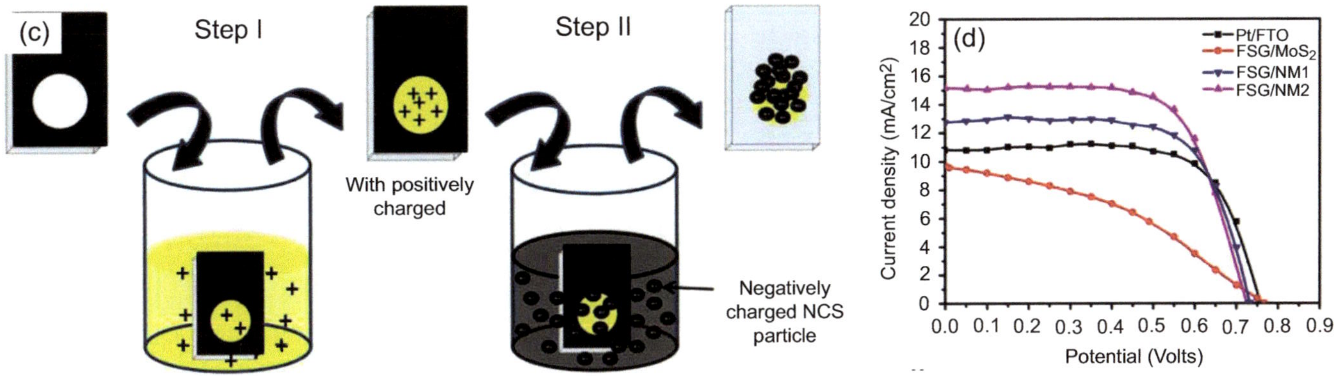

Figure 6.3 (a) Optimized bulk structure of NiS with rows of three-coordinated S ions exposed and direct dissociation of the I_2 molecule upon adsorption at the S-vacancy dimer with adsorption configuration, reprinted from Ref. [37], Copyright 2014, with permission from the Royal Society of Chemistry; (b) Nyquist plot with the structure diagram of NiS_2 and CoS_2/NiS_2 CEs, reprinted from Ref. [38], Copyright 2018, with permission from Elsevier; (c) Illustration of dip-coating process of NCS ($NiCo_2S_4$) CEs, reprinted from Ref. [41], Copyright 2016, with permission from Elsevier; (d) JV characteristics curve of NiMoS/FSG, reprinted from Ref. [42], Copyright 2018, with permission from Elsevier.

is drastic reduction in the cell impedance from 1000 to 43 ohms. The post-annealed films exhibit low resistance, thereby promoting the conductivity of the devices (7.44%) [36]. Apart from MoS_2 and WS_2, a number of sulphide-based CE materials are been reported by researchers. From the list of different sulphide-based CEs, let us have a view on one more material called nickel sulphide (NiS), since in the recent days Ni-based materials have played a prominent role in place of CEs. There are many reports available on NiS-based CEs in DSSCs. The information reported by Y. Li et al. will be more useful and interesting. In this article, the author discussed the growth of {0001} faceted single crystal NiS nanosheets via the hydrothermal approach and achieved a PCE of 8.62%, which is 17.1 times higher than that of the conventional CE. A brief molecular-level investigation into triiodide reduction reaction on the NiS(0001) surface was conducted by virtue of DFT calculations. The adsorption of I and activation of the I_2 molecule on the NiS (0001) surface supports the electrocatalytic activity of the NiS nanosheets. Bulk structure of NiS consists of six coordinated Ni^{2+} and S^{2-} as illustrated in Fig. 6.3a. The surface configuration of stable NiS(0001) is terminated by rows of three-coordinated sulfur ions (S3c) along the [0001] direction, from which one can see that on the stoichiometric NiS(0001) surface, Ni^{2+} cations retain the six-coordinated configuration and thus hardly adsorb I or I_2 species. To activate and adsorb the I_2 molecule, some defects such as S vacancies have to be introduced at the NiS(0001) surface, accompanied and determined by the catalyst preparation process. Owing to the presence of the S-vacancy in the NiS nanosheet (Fig. 6.3a), upon adsorption at the S-vacancy dimer in the direction of [0001], it is found that the I_2 molecule can readily dissociate into two I* atoms. The formed I* atom preferably sits at the S-vacancy and forms three Ni–I bonds at the distance, d(Ni–I), of B2.709 Å (Fig. 6.3a).

The outstanding electrocatalytic activity of the NiS(0001) surface can be expected, which results in high solar cell performance [37]. F. Li et al. fabricated NiS_2/CoS_2 composite by in situ chemical bath deposition and sulfurization process and used the electrode as CE in DSSCs. In this, CoS_2 nanoparticles cover the NiS_2, thereby reducing the sheet resistance and diffusion coefficient of CE materials. Moreover, the formed NiS_2/CoS_2 interface modulates the electronic structure of the composite, thereby contributing to tuning the performance of the device (8.22%), which is 24.36% higher than that of the conventional devices. Since the NiS_2 was grown vertically on the FTO, the interface/contact between them is very small as shown in Fig. 6.3b, leading to the increase in R_s. When the CoS_2 is loaded onto the NiS_2, the R_s value gets reduced by means of filling up the empty areas on the FTO (Fig. 6.3b) [38].

More recently, ternary/quaternary metal sulphides have attracted extensive attention as efficient CE materials due to their intrinsically excellent electrical conductivity and electrocatalytic activity originating from the unique coexistence of two different cations in a single crystal structure [39]. The properties of ternary and multicomponent are obtained from the optimized binary compounds. In the example of ternary NCS, the Ni^{2+} and Co^{3+} cations located at its tetrahedral and octahedral sites could provide not only richer redox reaction, but also more valence interchange or charge hopping between Ni and Co cations compared to the single phase of metal sulphides [40]. A.-L. Su et al. reported the synthesis of NCS particles via the solvothermal method, after which they are immobilized on FTO using a series of dip-coating processes as shown in Fig. 6.3c. DSSC based on the optimized NCS CE showed a higher PCE of 8.94%. Additionally, the NCS particles were successfully immobilized on the surface of flexible indium doped tin oxide/polyethylene naphthalate (ITO/PEN) substrates using the same solution approach. The DSSC with the flexible NCS CE can still exhibit a PCE of 8.62% (or 8.57% with the bended flexible NCS CE), which is higher than Pt CE (8.51%). More important, the proposed solution approach can also provide benefits such as easy fabrication process, low temperature, large scale, and no usage of any binders and expensive manufacturing tools [41]. R. Karthick et al. fabricated NiMoS on free-standing graphene (FSG) paper by electrodeposition method for FTO and Pt-free CE in DSSC. FSG along with metallic 2D nanoparticles (NiMoS) functions as an effective catalytic material, thereby allowing DSSCs to achieve enhanced efficiency of ~7% (Fig. 6.3d) [42]. Ternary researchers developed quaternary sulphide-based CEs such as Cu_2FeSnS_4, Cu_2CdSnS_4, and CZTS. K. Mokurala et al. demonstrated the annealing effect of quaternary chalcogenides (Cu_2FeSnS_4 and Cu_2CdSnS_4) in DSSC performance. The quaternary films were obtained by thermal decomposition of metal precursors, followed by sulfurization process. The device exhibits the PCE of 7.36%, which is most equivalent to that of Pt device [43].

6.5 Selenide-Based Electrode

As an important class of chalcogenides, semiconducting selenides have drawn enormous attention due to their distinctive electronic properties and interesting chemical behaviors. It has been investigated in a wide variety of potential applications, such as optical recording materials, sensor, laser materials, and photovoltaic fields [44–46]. Quite recently, metal selenides have emerged as a new class of CE electrocatalysts for DSSCs despite their conventional and successful utilization as sensitizers for photoelectrodes. Different types of binary selenides ($MoSe_2$, WSe_2, NiSe, CoSe, Bi_2Se_3, ZnSe , PdSe, etc.) and ternary selenides (NiCoSe, RuCoSe, etc.) act as electrocatalytic candidates and are employed to prepare CE for DSSCs. The performance of the solar cells featuring these selenides was compared with that of the conventional Pt CE.

6.5.1 Binary Selenides

Molybdenum and tungsten selenide ($MoSe_2$ and WSe_2) derivatives of chalcogenides possess layered structure where the chemical bond is formed within each layer (in-plane), while the weak out of plane van der Waals force exists across the layers [47, 48]. Due to 2D confinement of electrons, layered selenides ($MoSe_2$, WSe_2, etc.) exhibit unusual properties, particularly for catalytic applications. In this regard, L. T. L. Lee et al. proposed a few-layered $MoSe_2$ towards high catalytic activity in iodide/triiodide redox shuttles. The author demonstrated $MoSe_2$/Mo substrate for CEs. Initially Mo is sputtered over the soda lime glass and then selenization of chalcogenides takes place on placing Mo substrate and selenium pellets in the quartz tube furnace. The schematic representation of the experimental setup is shown in Fig. 6.4a. Upon selenization, the expansion of volume takes place, contributing towards few-layered $MoSe_2$ (Fig. 6.4a). The few-layered $MoSe_2$ displays high catalytic efficiency for the regeneration of I_2 species, which in turn yields a photovoltaic energy conversion efficiency of 9.00% with R_{ct} of 1.39 Ω/cm^2, while the identical photoanode coupling with "champion" electrode based on Pt nanoparticles on FTO glass generates efficiency of only 8.68% with R_{ct} of 0.22 Ω/cm^2. Even though the charge transfer resistance value is high for $MoSe_2$/Mo CEs, the enhanced performance of

$MoSe_2$ is mainly attributed to the reduced sheet resistance value (1.74 Ω/cm^2) of $MoSe_2$/Mo, while for the Pt it is (12.52 Ω/cm^2). The author has also made an initial attempt to fabricate large-scale crack-free selenide CEs with an area of 100×10 mm^2 [49]. Surprisingly during the same period of time, H. Chen et al. also fabricated the combo of $MoSe_2$/Mo CEs, where $MoSe_2$ layer functions as the electrocatalytic centre and the base Mo layer acts as the conductive collector for electron extraction with a PCE of 8.13%. To add strength to the statement, the electrochemical results (Fig. 6.4b) reported by the author was witnessing one, where the obtained sheet resistance value of $MoSe_2$/Mo (2.64 Ω/cm^2) was seven times lesser than that of Pt (15.98 Ω/cm^2) [50]. X. Yuan et al. fabricated hierarchical $MoSe_2$ nanoflowers (Fig. 6.4c) via hydrothermal approach and used as an efficient CE. The hierarchal architecture of nanoflowers with multilayer speeds up the process of reduction of electrolyte and enhances the electron transfer efficiency. $MoSe_2$ nanoflower CE exhibits an impressive PCE of 7.01% and outperforms the standard Pt (6.38%) [51].

J. Guo et al. designed and synthesized three chalcogenides ($MoSe_2$, WSe_2, and $TaSe_2$) with similar crystal structures by the reductant-free solvothermal route. The author systematically explained the concept of catalytic activity of all the three chalcogenides from both the experimental and the theoretical method. The performance of WSe_2 (7.48%) and $TaSe_2$ (7.32%) was higher than that of $MoSe_2$ (6.70%), the reason behind the enrichment in efficiency is due to high catalytic active sites of the materials. The structures of I_3 and its adsorption on $MoSe_2$(001) and WSe_2(001) surfaces, as well as the corresponding projected density of states (PDOS), are shown in Fig. 6.5a. The results indicate that the I_3 complex is preferentially adsorbed at the top of Se on the $MoSe_2$(001) surface and the hollow of the WSe_2(001) surface. Bader charge analysis suggests that the I_3 complex gains approximately 0.14e from the WSe_2(001) surface but only 0.11e from the $MoSe_2$(001) surface. Thus, the WSe_2 is more efficient in regenerating I_3^-/I^- redox couples than $MoSe_2$.The theoretical calculation results agree well with the experimental data for the two CEs. The authors further analyzed the PDOS) diagram for I_3^- ($MoSe_2$ and WSe_2) in the configuration. From the PDOS diagram, it is clear that the I 5p orbital, W and Mo d orbitals, and the Se 4p orbital exist in obvious hybridization, indicating the relatively large binding energy of the I_3 complex adsorbed onto $WSe_2(001)$ or $MoSe_2(001)$ [52].

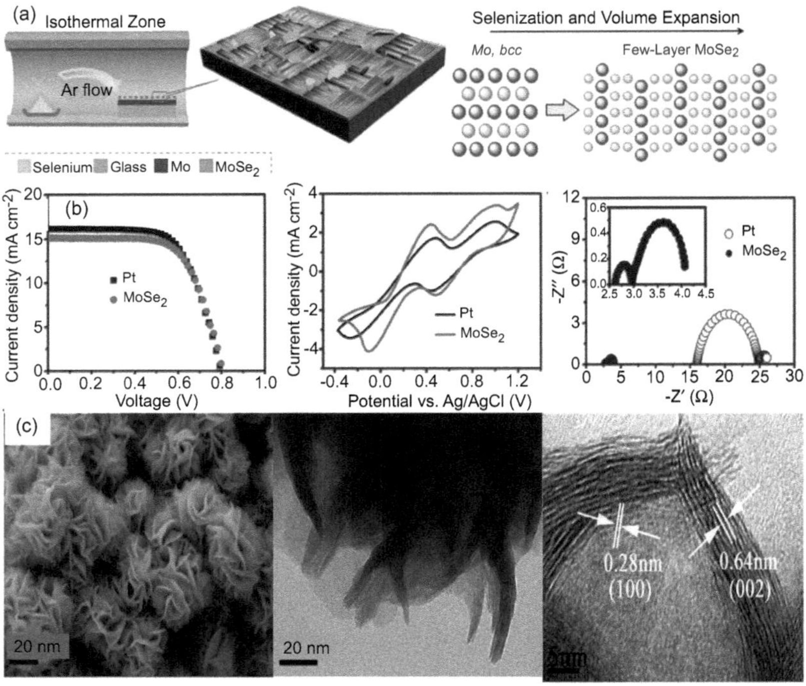

Figure 6.4 (a) Schematic representation of selenization of Mo in furnace tube with the formation of few-layer $MoSe_2$ from body-centered cubic Mo, where the volume expansion occurs, adapted from Ref. [49] under Creative Commons License; (b) JV, CV, Nyquist plot of MoSe/Mo CE, reprinted from Ref. [50], Copyright 2014, with permission from the Royal Society of Chemistry; and (c) Morphology of MoSe nanoflowers, reprinted from Ref. [51], Copyright 2018, with permission from Elsevier.

Zhou et al. reported regulation of microstructure and composition of cobalt selenide (CoSe) by the electrochemical atomic layer deposition (ECALD) method for first time with a PCE of 9.28%. In this article, the deposition potential and time of Se and Co along with varying composition (0.8, 1, and 1.2) were carried out using constant potential square wave method in an automatic ECALD system. The schematic representation of ECALD setup is shown in Fig. 6.5b [53]. By varying the potential of Se and keeping the potential of Co constant, the composition of deposits could be adjusted to different ratios of Co and Se. The author also stated that the obtained efficiency is 136% higher than that of the conventional Pt CE (6.84%). The reason behind the increase in PCE is mainly attributed to the more active catalytic site and less charge transfer resistance, which was confirmed by electrochemical studies. From the CV analysis, the larger peak current density and the lower value of E_{pp} revealed a higher catalytic activity of the CoSe CE and the obtained CV is shown in Fig. 6.5b [54].

Researchers also paid their attention to replacing the transparent conducting substrate used for fabricating CEs to reduce the fabrication cost of DSSCs further. In this connection, C. Bao et al. deposited nickel selenide (NiSe) on nickel (Ni) foils, thereby replacing both Pt and FTO substrate. The Ni foils acting as the base substrate helps in the reaction process and enhances the electron transport. Combining the effects of both metallic conducting Ni and electrocatalytic NiSe (NiSe/Ni) CEs delivers PCE of 6.75% with an excellent electrochemical stability [55]. Applying the similar way of approach, L. Chen et al. fabricated selenide nanosheets on metal fibers and used as CE. However, it is difficult for the electrocatalyst to form good quality films on fiber electrodes with traditional methods such as drop casting and spin coating. Therefore, Chen and coauthors realized $Ni_{0.85}Se$ and $Co_{0.85}Se$ nanosheet on Ni and Co fibers through a one-step solvothermal reaction for construction of fiber DSSCs (Fig. 6.5c).

The selenide nanosheets formed uniform films with strong adhesion to fibers as a result of the in situ growth. The fiber Ni–$Ni_{0.85}Se$ and Co–$Co_{0.85}Se$ nanosheet CEs exhibit higher catalytic activity than the fiber Pt CE, and consequently the two fiber metal selenide CEs yield PCEs comparable to that of the fiber Pt CE (Fig. 6.5c) [56]. In a study on TiOSe as CE, C.-T. Li et al. fabricated hierarchical $TiO_{1.1}Se_{0.9}$ nanospheres and nanorod core/shell composites get wrapped with the carbon cloth (Fig. 6.5d). Thus, each carbon fiber with its $TiO_{1.1}Se_{0.9}$ layer has a hierarchical core–shell structure, in which the carbon fiber acts as a one-dimensional conductive core for electron transfer and the $TiO_{1.1}Se_{0.9}$ layer around the fiber functions as an electrocatalytic shell. Pt and FTO free carbon cloth wrapped TiOSe CE delivers an impressive efficiency of 10.47% at 0.5 sun condition [57].

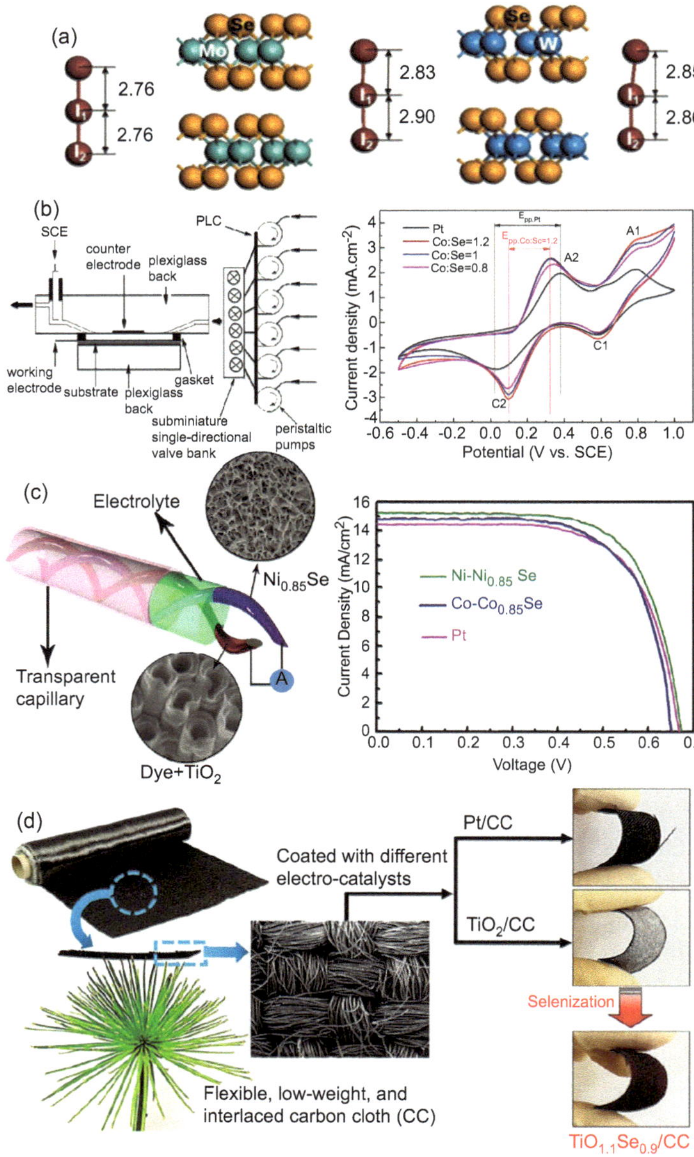

Figure 6.5 (a) Side views of I_3 complex adsorption on 001 surfaces ($MoSe_2$ and WSe_2) with an optimized I_3 complex and PDOS for I_3 complex adsorption, reprinted from Ref. [52], Copyright 2015, with permission from the Royal Society of Chemistry; (b) Automated ECALD setup for depositing CoSe and its corresponding CV, reprinted from Ref. [53], Copyright 2005, and Ref. [54],

Copyright 2016, with permission from Elsevier; (c) TCO-free fiber DSSC and its corresponding JV curve of Ni–NiSe and Co–CoSe, reprinted from Ref. [56], Copyright 2016, with permission from the Royal Society of Chemistry; (d) Images of carbon cloth, which is a foldable (top left) and low-weight (bottom left) substrate formed of many interlaced bundles of carbon fibers (middle), the photos of the electrodes of Pt/CC, TiO_2/CC, and $TiO_{1.1}Se_{0.9}$/CC (right), reprinted from Ref. [57], Copyright 2017, with permission from the Royal Society of Chemistry.

6.5.2 Ternary/Quaternary/Penternary Selenides

Since materialists are always keen on developing novel materials, compounds, and composites which leads in developing different materials into single compound (i.e., ternary, quaternary, penternary etc). These kinds of materials improve the performance by means of combined properties of all the materials. First, we consider some of the examples of ternary-based selenides as CEs in the fabrication of DSSCs, followed by quaternary and penternary selenides. Y. Zhou et al. demonstrated electrodeposition of ruthenium cobalt selenide (RuCoSe) as ternary alloy for Pt-free counter electrode in the fabrication of DSSCs. For comparative purposes, the author also fabricated CoSe, RuSe, and RuCoSe CEs. The resultant solar cells generated increased efficiencies of 7.53%, 7.81%, and 9.07%, respectively (Fig. 6.6a), benefiting from increased catalytic activity of metal selenide alloy CEs. Electrodeposited RuCoSe ternary alloy yields a PCE of 9.07%, which is higher than that of Pt tailored devices (7.03%) [58]. Electrochemical characterizations including CV and EIS are used to evaluate electrodic reaction processes. Two pairs of redox peaks are recorded for all CV curves (Fig. 6.6b), indicating a two-electron reaction in I_3^- to I^- interconversion. $I_3^- + 2e^- = 3I^-$ is the major process at the CE surface; therefore, J_{Red1}, E_{pp}, and τ are crucial parameters in controlling the total catalytic reaction kinetics. In comparison with expensive Pt CE, an increased J_{Red1} and reduced E_{pp} suggest that more electrons participate in the I_3^- reduction process with less energy requirement. The easy preparation method and promising results provide a new method for decreasing cost with enhancing photovoltaic performance of DSSCs.

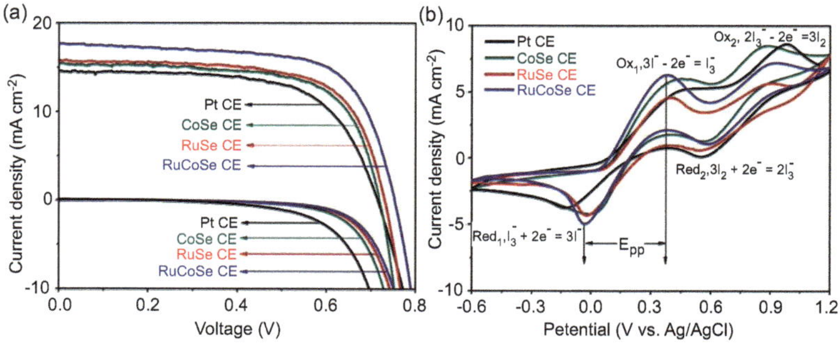

Figure 6.6 (a) JV characteristic curve and (b) CV graphs of Co-, Ru-, RuCo-based Se CEs. Reprinted from Ref. [58], Copyright 2018, with permission from Elsevier.

J. Theerthagiri et al. fabricated ternary selenides $(Ni_{1-x}Co_x)Se_2$ as the electrocatalyst for triiodide reduction in DSSCs using the one-step hydrothermal reduction route. The author reported $(Ni_{1-x}Co_x)Se_2$ with different compositions (x = 0.0, 0.1, 0.5, and 1.0). A PCE of 6.02% is achieved for the ratio of $Ni_{0.5}Co_{0.5}Se_2$, which is equivalent to that of Pt-based CEs (6.11%) [59].

After the successful formation of ternary-based selenides, researchers started moving towards quaternary and penternary selenide compounds. Y.-F. Du et al. demonstrated the one-step solvothermal synthesis method for fabricating $Cu_2ZnSnSe_4$ in stoichiometric ratio without using any surfactants or templates. The author reported the variation of film thickness and the best PCE of 3.85% [60]. A similar material $Cu_2ZnSnSe_4$ was demonstrated by S.-J. Yuan et al., with a PCE of 7.82%. The author synthesized quaternary chalcogenide compounds via solution based in situ synthesis [61]. F. Ozel et al. fabricated penternary chalcogenide [$Cu_2CoSn(SeS)_4$ and $Cu_2ZnSn(SeS)_4$] nanocrystals as effective catalytic materials for fabricating CEs in DSSCs. The synthesized penternary chalcogenides hold single crystalline kesterite phase with an adequate stoichiometric ratio and bandgap energy of 1.18–1.45 eV. The photovoltaic results demonstrated that DSSC with a $Cu_2CoSn(SeS)_4$ nanocrystal-based CE achieved the best PCE of 6.47%, which is higher than the same photoanode employing $Cu_2ZnSn(SeS)_4$ nanocrystals (3.18%) and Pt (5.41%) CEs. These promising results highlight the potential application of penternary chalcogen $Cu_2CoSn(SeS)_4$ nanocrystals in low-cost, high-efficiency, Pt-free DSSCs. The reported JV and electrochemical studies are shown in Fig. 6.7 [62].

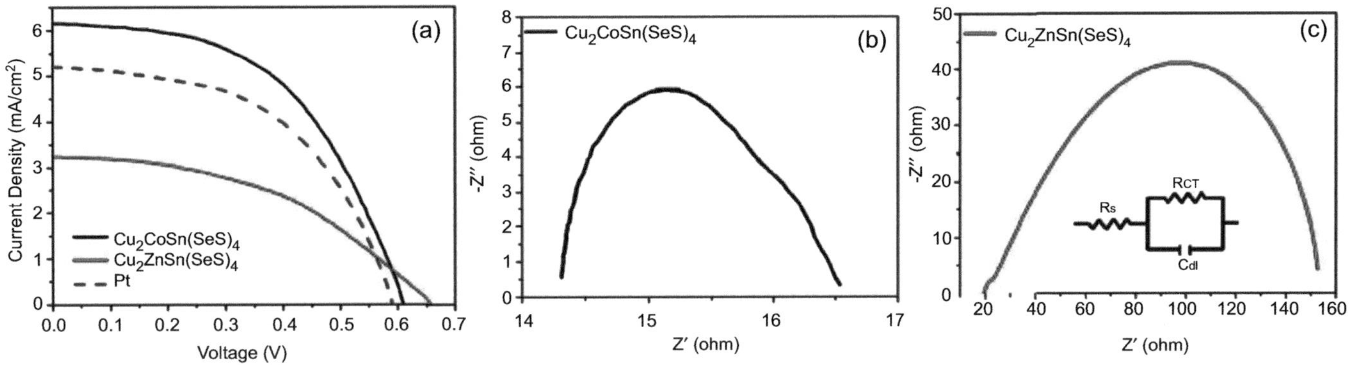

Figure 6.7 (a) JV characteristics of $Cu_2CoSn(SeS)_4$ and $Cu_2ZnSn(SeS)_4$ nanocrystals. Nyquist plot of (b) $Cu_2CoSn(SeS)_4$ nanocrystals and (c) $Cu_2ZnSn(SeS)_4$ nanocrystals. Adapted from Ref. [62] under Creative Commons License.

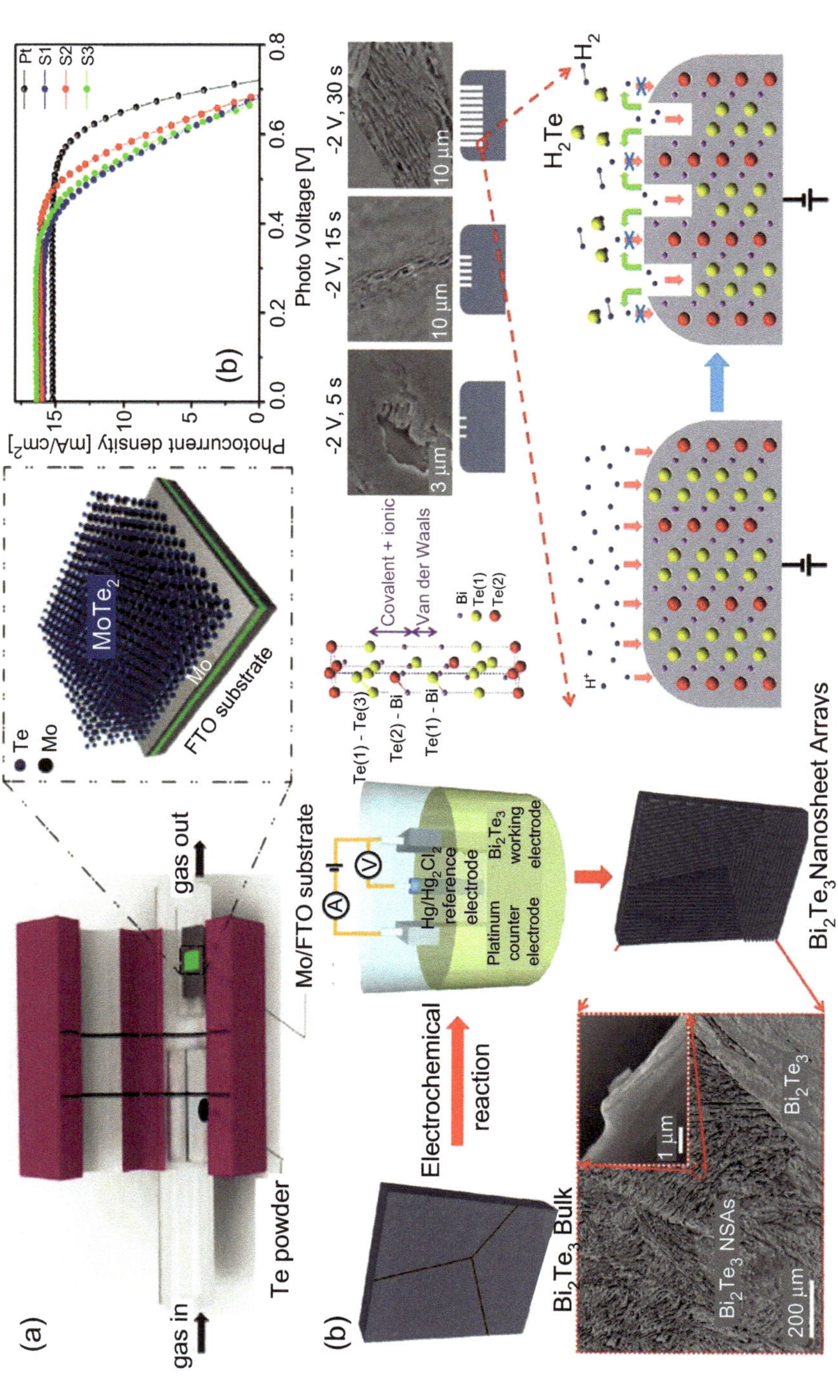
(a)
gas in
Te powder
gas out
Mo/FTO substrate
Te
Mo
MoTe2
Mo
FTO substrate
(b)
Photocurrent density [mA/cm2]
Photo Voltage [V]
Pt
S1
S2
S3
(b)
Bi2Te3 Bulk
Electrochemical reaction
Platinum counter electrode
Hg/Hg2Cl2 reference electrode
Bi2Te3 working electrode
A
V
Bi2Te3 Nanosheet Arrays
Bi2Te3 NSAs
Bi2Te3
200 μm
1 μm
Te(1) - Te(3)
Te(2) - Bi
Te(1) - Bi
Covalent + ionic
Van der Waals
Bi
Te(1)
Te(2)
-2 V, 5 s
-2 V, 15 s
-2 V, 30 s
3 μm
10 μm
10 μm
H+
H2Te
H2

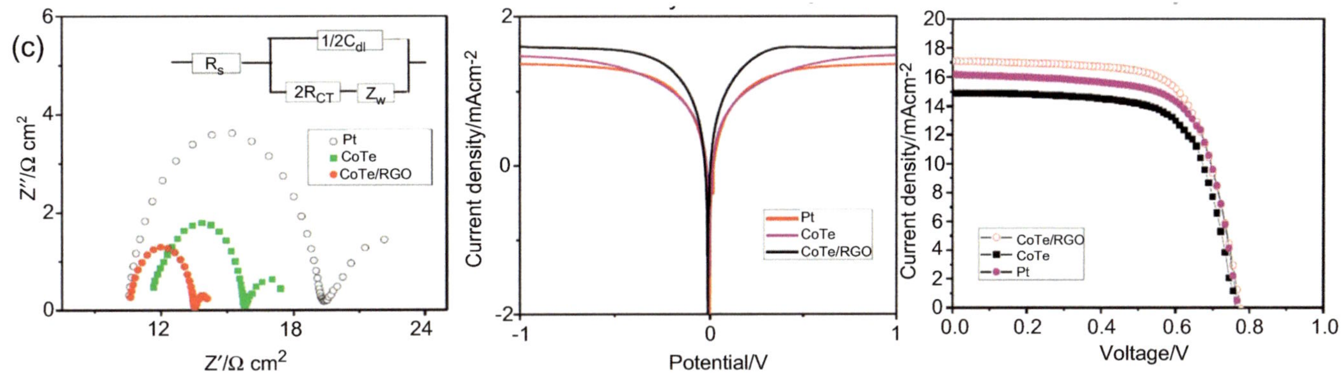

Figure 6.8 (a) Illustration of two-zone chemical vapor deposition chamber for the preparation of $MoTe_2$ from Mo/FTO substrate with its corresponding JV curve, adapted from Ref. [67] under Creative Commons License.; (b) Schematic representation of electrolysis process for fabricating Bi_2Te_3 with its mechanism at different steps, reprinted from Ref. [68], Copyright 2014, with permission from the Royal Society of Chemistry.; (c) Electrochemical and JV characteristics of Pt, CoTe, and CoTe/rGO CEs, reprinted from Ref. [71], Copyright 2015, with permission from Elsevier.

6.6 Tellurium-Based Electrode

One more interesting material in the family of chalcogenides is telluride (Te). Recently, focus has been placed on telluride-based materials such as $MoTe_2$, WTe_2, and $BiTe_2$ in the field of electronics and optoelectronic devices [63, 64]. Considering the case of $MoTe_2$, it also exhibits characteristics like other chalcogenide-based materials. It has the lowest direct bandgap of –1.0 eV, which is the lowest value in the family of chalcogenide materials. A single layer of $MoTe_2$ has a direct bandgap of 1.1 eV, which is most equivalent to the bandgap value of silicon (1.1 eV) [65]. The crystal structure of $MoTe_2$ is highly stable in semiconducting (2H) and metallic (1T) phase in nature [66]. By means of its layered form of structure and its porosity in nature, it tends to play a role in electrocatalytic activity, thereby finding its path in the fabrication of DSSCs as counter electrodes. There are very few reports available on telluride-based counter electrodes.

S. Hussain et al. in 2018 reported fabrication of large area MoTe thin films via sputtering combined with post-annealing treatment using chemical vapor deposition. The scheme of depositing MoTe on the FTO conducting substrate using two zone chemical vapor deposition chambers is shown in Fig. 6.8a. The author assessed the effect of thickness of the film by varying the deposition rate of time (20, 30, 40 min) in sputtering process and named S1, S2, S3, respectively. An optimized thickness of 30 min (S2) sample in the fabrication of DSSC as the counter electrode delivers a high PCE of 7.25%, which is one of the best PCE values for the telluride-based counter electrode in DSSCs (Fig. 6.8a). The charge transfer kinetics and the internal resistance of DSSCs were reported by measuring EIS using symmetrical cell (CE/electrolyte/CE) setup [67].

Similarly, H.-W. Tsai et al. reported fabrication of large-scale single crystal bismuth telluride (Bi_2Te_3) nanosheet array for counter electrode using single-step electrolytic process and its scheme is given in Fig. 6.8b [68]. Bi_2Te_3 is one of the chalcogenide-based thermoelectric materials that helps in reducing the thermal conductivity during the energy conversion. In this regard, Bi_2Te_3 comes with an excellent figure of merit at room temperature with high thermoelectric effect, namely, ZT value, which is used to examine the thermoelectric performance of materials [69]. In addition to the figure of merit, the materials also possess some interesting physical properties like high surface area, which helps in boosting the electrical conductivity of devices [70].

The author fabricated Bi_2Te_3 nanosheets in a three-electrode system using polished bulk Bi_2Te_3 as working electrode, platinum as counter electrode, and Hg/Hg_2Cl_2 as reference electrode. Before carrying out this experiment, bulk Bi_2Te_3 sample was prepared by taking Bi and Te ingots in a sealed evacuated quartz tube and annealing at 850°C for 48 h using furnace. The fabricated Bi_2Te_3 nanosheets arrays deliver PCE of 1.12% with J_{sc} of 6.89 mA/cm^2, V_{oc} of 0.58, and FF of 0.28 [68].

J. Jia et al. fabricated telluride-based composite for counter electrode using two-step synthesis hydrothermal and hummer's method. The author demonstrated hybrid cobalt telluride/reduced graphene oxide (CoTe/rGO) for fabricating counter electrode and achieved the best PCE of 9.17%, which is higher than that of the conventional counter electrode (Pt ~ 8.17%). The fabricated counter electrode shows low resistance and prominent catalytic activity for reduction of electrolyte (I_3^- ions). The author also stated that the enhanced efficiency of hybrid CoTe/rGO was mainly attributed to its high surface area than the CoTe. Nyquist, Tafel, and JV curves obtained for bare Pt, CoTe, and CoTe/rGO was shown in Fig. 6.8c [71].

Table 6.1 presents the summary of methodologies involved in fabricating the chalcogenide- (S, Se, and Te) based materials with their composites and the photovoltaic properties of the corresponding chalcogenides.

Table 6.1 Summarized photovoltaic properties of chalcogenide-based CEs in DSSCs

CE	Methodology	J_{sc} (mA/cm^2)	V_{oc} (V)	FF	PCE (%)	Ref.
MoS_2	Wet chemical	13.84	0.76	0.73	7.59	[29]
WS_2	Wet chemical	14.13	0.78	0.70	7.73	[29]
MoS_2/graphene	Electro-chemical	16.96	0.72	0.65	8.01	[31]
WO_x@WS_2@carbon	Vaporization	15.48	0.72	0.69	7.71	[35]

(*Continued*)

Table 6.1 (*Continued*)

CE	Methodology	J_{sc} (mA/cm^2)	V_{oc} (V)	FF	PCE (%)	Ref.
WS_2–W_5O_{14}–MWCNTs	Hydrothermal	15.19	0.74	0.65	7.44	[36]
NiS	Hydrothermal	18.71	0.73	0.62	8.62	[37]
NiS_2/CoS_2	Chemical bath deposition	17.45	0.72	0.65	8.22	[38]
$NiCo_2S_4$	Dip coating	17.17	0.74	0.69	8.89	[41]
NiMoS/FSG	Electro-chemical	15.0	0.72	0.64	7.0	[42]
Cu_2FeSnS_4	Thermal decomposition	17.37	0.70	0.57	7.36	[43]
Cu_2CdSnS_4	Thermal decomposition	16.38	0.69	0.56	7.12	[43]
$MoSe_2$/Mo	Sputtering/selenization	15.07	0.80	0.67	8.13	[50]
WSe_2	Solvothermal	15.50	0.73	0.66	7.48	[52]
$TaSe_2$	Solvothermal	15.81	0.73	0.64	7.32	[52]
CoSe	ECALD	18.56	0.78	0.64	9.28	[53]
NiSe/Ni	Solvothermal	15.90	0.62	0.69	6.75	[55]
$TiO_{1.1}Se_{0.9}$/CC	Dip coating	17.22	0.79	0.70	9.47	[57]
RuCoSe	Electro-deposition	17.86	0.75	0.67	9.07	[58]
$(Ni_{1-x}Co_x)Se_2$	Hydrothermal	15.65	0.70	0.54	6.02	[59]
$Cu_2ZnSnSe_4$	Drop casting	15.54	0.78	0.65	7.82	[61]
$Cu_2CoSn(SeS)_4$	Hot injection	6.14	0.61	0.57	6.47	[62]
$Cu_2ZnSn(SeS)_4$	Hot injection	3.23	0.65	0.45	3.18	[62]
MoTe	Sputtering/CVD	16.0	0.69	0.65	7.25	[67]
CoTe	Hydrothermal	14.89	0.76	0.68	7.75	[71]
CoTe/rGO	Hydrothermal	17.41	0.77	0.68	9.18	[71]

6.7 Future Scope and Challenges

With all these promising properties and significant progress in chalcogenides, their commercial applications still seem to be a long way off. To apply them in pilot scale, some critical challenges have yet to be overcome. Despite their unique properties compared to other semiconductors, they usually suffer from irreversibility, poor rate capability, and roundtrip energy inefficiency. Moreover, compatibility with electrolyte is a major concern; the formation of soluble polysulphides in the electrolyte and its effect on electrochemical characterizations lead to diminishing performance of devices. These obstacles are deeply rooted in transferring the technology into commercial applications and cannot be simply addressed without a comprehensive understanding of their origin and effects.

6.8 Conclusion

This chapter focuses on the importance of counter electrodes and the application of chalcogenides- (S, Se, Te) based binary, ternary, quaternary, penternary materials as counter electrode in the fabrication of DSSCs. The methodology and electrochemical and photovoltaic properties of fabricated DSSCs employing chalcogenide-based CEs are summarized and discussed. The data presented indicate that chalcogenide-based CEs will surpass the conventional Pt CE in the fabrication of DSSCs. The challenges faced in transforming the device into technology discussed in this chapter are more useful for the budding researchers to find solutions. Furthermore, the large family of chalcogenides, such as MoS_2, $MoSe_2$, $MoTe_2$, WS_2, WSe_2, WTe_2, $FeSe_2$, TaS_2, and $NbSe_2$, should also be explored with other inorganic and organic materials. The family of chalcogenides is enormously large and expected to play a vital role in developing low-cost highly efficient DSSCs for commercial applications.

Acknowledgment

The author A. Arulraj expresses his gratitude to Dr. V. Sivakumar, Professor, Department of Physics, University College of Engineering, BIT Campus, Anna University, Tiruchirappalli and Dr. B. Subramanian, Principal Scientist, Functional Materials Division, CSIR-Central Electrochemical Research Institute (CECRI), Karaikudi for their constant support. The author Dr. M. Ramesh gratefully acknowledges the Department of Science and Technology (DST) for granting INSPIRE faculty award (DST/INSPIRE/04/2015/002860).

References

1. A. Hagfeldt, G. Boschloo, L. Sun and H. Petterson, Dye-sensitized solar cells, *Chemical Reviews*, 110, 2010, 695–663.
2. Q. Zhang, K. Park, J. Xi, D. Myers and G. Gao, Recent progress in dye-sensitized solar cells using nanocrystalline aggregates, *Advanced Energy Materials*, 1, 2011, 988–1001.
3. H. Choi, C. Nahm, J. Kim, C. Kim, S. Kang, T. Hwang and B. Park, Review paper: toward highly efficient quantum-dot and dye-sensitized solar cells, *Current Applied Physics*, 13, 2013, S2–S13.
4. H. Gerischer, M. E. Michel-Beyerle, F. Rebentrost and H. Tributsch, Sensitization of charge into semiconductors with large band gap, *Electrochimica Acta*, 13, 1968, 1509–1515.
5. H. Tributsch and H. Gerischer, The use of semiconductor electrodes in the study of photochemical reactions, *Berichte der Bunsengesellschaft Physical Chemistry*, 73, 1969, 850–854.
6. H. Tributsch and M. Calvin, Electrochemistry of excited molecules: photoelectrochemical reactions of chlorophyll, *Photochemistry and Photobiology*, 14, 1971, 95–112.
7. W. D. K. Clark and N. Sutin, Spectral sensitization of n-type titanium dioxide electrodes by polypyridine ruthenium(II) complexes, *Journal of American Chemical Society*, 99, 1977, 4676–4682.
8. M. T. Spitler and B. A. Parkinson, Dye sensitization of single crystal semiconductor electrodes, *Accounts of Chemical Research*, 42, 2009, 2017–2029.
9. B. O'Regan and M. Gratzel, A low-cost high-efficiency solar cell based on dye-sensitized colloidal TiO_2 films, *Nature*, 353, 1991, 737–740.

10. M. Gratzel, Recent advances in sensitized mesoscopic solar cells, *Accounts of Chemical Research*, 42, 2009, 1788–1798.

11. X. Cui, J. Xiao, Y. Wu, P. Du, R. Si, H. Yang, H. Tian, J. Li, W.-H. Zhang, D. Deng and X. Bao, A graphene composite material with single cobalt active sites: a highly efficient counter electrode for dye-sensitized solar cells, *Angewandte Chemie International Edition*, 55, 2016, 6707–6711.

12. C. T. Lee, J. D. Peng, C. T. Li, Y. L. Tsai, R. Vittal and K. C. Ho, Ni_3Se_4 hollow architectures as catalytic materials for the counter electrodes of dye-sensitized solar cells, *Nano Energy*, 10, 2014, 201–211.

13. S. Z. Butler, S. M. Hollen, L. Cao, Y. Cui, J. A. Gupta, H. R. Gutierrez, T. F. Heinz, S. S. Hong, J. Huang, A. F. Ismach, E. Johnston-Halperin, V. V. Plashnitsa, R. D. Robinson, R. S. Ruoff, S. Salahuddin, J. Shan, L. Shi, M. G. Spencer, M. Terrones, W. Windl and J. E. Goldberger, Progress, challenges, and opportunities in two-dimensional materials beyond graphene, *ACS Nano*, 7, 2013, 2898–2926.

14. X. Duan, C. Wang, A. Pan, R. Yu and X. Duan, Two-dimensional transition metal dichalcogenides as atomically thin semiconductors: opportunities and challenges, *Chemical Society Reviews*, 44, 2015, 8859–8876.

15. D. Jariwala, V. K. Sangwan, L. J. Lauhon, T. J. Marks and M. C. Hersam, Emerging device applications for semiconducting two-dimensional transition metal dichalcogenides, *ACS Nano*, 8, 2014, 1102–1120.

16. S. Park, H. Boo and T. D. Chung, Electrochemical non-enzymatic glucose sensors, *Analytica Chimica Acta*, 556, 2006, 46–57.

17. C. M. Chen, C.-H. Chen and T. C. Wei, Chemical deposition of platinum on metallic sheets as counter electrodes for dye-sensitized solar cells, *Electrochimica Acta*, 55, 2010, 1687–1695.

18. X. Yin, Z. Xue and B. Liu, Electrophoretic deposition of Pt nanoparticles on plastic substrates as counter electrodes in dye-sensitized solar cells, *Journal of Power Sources*, 196, 2011, 2422–2426.

19. T. N. Murakami and M. Gratzel, Counter electrodes for DSC: application of functional materials as catalysts, *Inorganica Chimica Acta*, 361, 2008, 572–580.

20. W. Choi, N. Choudhary, G. H. Han, J. Park, D. Akinwande and Y. H. Lee, Recent development of two-dimensional transition metal dichalcogenides and their applications, *Materials Today*, 20, 2017, 116–130.

21. D. H. Keum, S. Cho, J. H. Kim, D.-H. Choe, H.-J. Sung, M. Kan, H. Kang, J.-Y. Hwang, S. W. Kim, H. Yang, K. J. Chang and Y. H. Lee, Band gap opening in few-layered monoclinic $MoTe_2$, *Nature Physics*, 11, 2015, 482–486.

22. M. N. Ali, J. Xiong, S. Flynn, J. Tao, L. M. Schoop, T. Liang, N. Haldolaarachchige, M. Hirschberger, N. P. Ong and R. J. Cava, Large, non-saturating magnetoresistance in WTe_2, *Nature*, 514, 2014, 205–208.

23. F. Xia, H. Wang, D. Xiao, M. Dubey and A. Ramasubramanian, Two-dimensional material nanophotonics, *Nature Photonics*, 8, 2014, 899–907.

24. M. Pumera, Z. Sofer and A. Ambrosi, Layered transition metal dichalcogenides for electrochemical energy generation and storage, *Journal of Materials Chemistry A*, 2, 2014, 8981– 8987.

25. F. Gustavsson, F. Svahn, U. Bexell and S. Jacobson, Nanoparticle based and sputtered WS_2 low-friction coatings – differences and similarities with respect to friction mechanisms and tribofilm formation, *Surface and Coatings Technology*, 232, 2013, 616–626.

26. A. Sobczynski, Molybdenum disulfide as a hydrogen evolution catalyst for water photodecomposition on semiconductors, *Journal of Catalysis*, 131, 1991, 156–166.

27. A. B. Laursen, S. Kegnæs, S. Dahl and I. Chorkendorff, Molybdenum sulphides: efficient and viable materials for electro- and photoelectrocatalytic hydrogen evolution, *Energy & Environmental Science* 5, 2012, 5577–5591.

28. J. W. Suk, R. D. Piner, J. An and R. S. Ruoff, Mechanical properties of monolayer graphene oxide, *ACS Nano*, 4, 2010, 6557–6564.

29. M. Wu, Y. Wang, X. Lin, N. Yu, L. Wang, L. Wang, A. Hagfeldt and T. Ma, Economical and effective sulphide catalysts for dye-sensitized solar cells as counter electrodes, *Physical Chemistry Chemical Physics*, 13, 2011, 19298–19301.

30. V. H. V. Quy, E. Vijayakumar, P. Ho, J.-H. Park, J. A. Rajesh, J. Kwon, J. Chae, J.-H. Kim, S.-H. Kang and K.-S. Ahn, Electrodeposited MoS_2 as electrocatalytic counter electrode for quantum dot and dye-sensitized solar cells, *Electrochimica Acta*, 260, 2018, 716–725.

31. S. Li, H. Min, F. Xu, L. Tong, J. Chen, C. Zhu and L. Sun, All electrochemical fabrication of MoS_2/graphene counter electrodes for efficient dye-sensitized solar cells, *RSC Advances*, 6, 2016, 34546–34552.

32. A. Arulraj, M. Ramesh, B. Subramanian and G. Senguttuvan, In-situ temperature and thickness control grown 2D-MoS_2 via pulsed laser ablation for photovoltaic devices, *Solar Energy*, 174, 2018, 286–295.

33. D. Merki and X. Hu, Recent developments of molybdenum and tungsten sulphides as hydrogen evolution catalysts, *Energy and Environmental Science*, 4, 2011, 3878–3888.

34. D. Voiry, H. Yamaguchi, J. Li, R. Sliva, D. C. Alves, T. Fujita, M. Chen, T. Asefa, V. B. Shenoy, G. Eda and M. Chhowalla, Enhanced catalytic activity in strained chemically exfoliated WS_2 nanosheet for hydrogen evolution, *Nature Materials*, 12, 2013, 850–855.

35. Z. Shen, M. Wang, L. Liu, M. V. Sofianos, H. Yang, S. Wang and S. Li, Carbon-coated three-dimensional WS_2 film consisting of WO_3@ WS_2 core-shell blocks and layered WS_2 nanostructures as counter electrodes for efficient dye-sensitized solar cells, *Electrochimica Acta*, 266, 2018, 130–138.

36. L. Keawphaisan, V. Harnchana, S. Pimanpang and V. Amornkitburung, Hydrothermal synthesis of the composited WS_2-W_5O_{14}-MWCNTs for high performance dye-sensitized solar cells, *Journal of Materials Science: Materials in Electronics*, 28, 2017, 18765–18772.

37. Y. Li, H. Wang, H. Zhang, P. Liu, Y. Wang, W. Fang, H. Yang, Y. Li and H. Zhao, A {0001} faceted single crystal NiS nanosheets electrocatalyst for dye-sensitized solar cells: sulphur-vacancy induced electrocatalytic activity, *Chemical Communications*, 50, 2014, 5569–5571.

38. F. Li, J. Wang, L. Zheng, Y. Zhao, N. Huang, P. Sun, L. Fang, L. Wang and X. Sun, In situ preparation of NiS_2/CoS_2 composite electrocatalytic materials on conductive glass substrates with electronic modulation for high-performance counter electrodes of dye-sensitized solar cells, *Journal of Power Sources*, 384, 2018, 1–9.

39. J. Xiao, L. Wan, S. Yang, F. Xiao and S. Wang, Design hierarchical electrodes with highly conductive $NiCo_2S_4$ nanotube arrays grown on carbon fiber paper for high-performance pseudocapacitors, *Nano Letters*, 14, 2014, 831–838.

40. H. Chen, J. Jiang, L. Zhang, H. Wan, T. Qi and D. Xia, Highly conductive $NiCo_2S_4$ urchin-like nanostructures for high-rate pseudocapacitors, *Nanoscale*, 5, 2013, 8879–8883.

41. A.-L. Su, M.-N. Lu, C.-Y. Chang, T.-C. Wei and J.-Y. Lin, Scalable fabrication of efficient $NiCo_2S_4$ counter electrodes for dye-sensitized solar cells using a facile solution approach, *Electrochimica Acta*, 222, 2016, 1410–1416.

42. R. Karthick, A. Arulraj, M. Ramesh and M. Selvaraj, Free-standing graphene/NiMoS paper as cathode for quasi-solid state dye-sensitized solar cells, *Journal of Colloids and Interface Science*, 530, 2018, 179–188.

43. K. Mokurala and S. Mallick, Effect of annealing atmosphere on quaternary chalcogenides-based counter electrodes in dye-sensitized solar cells performance: synthesis of Cu_2FeSnS_4 and Cu_2CdSnS_4 nanoparticles by thermal decomposition process, *RSC Advances*, 7, 2017, 15139–15148.

44. H. Ahari, C. L. Bowes, T. Jiang, A. Lough, G. A. Ozin, R. L. Bedard, S. Petrov and D. Young, Nanoporous tin(IV) chalcogenides: flexible open-framework nanomaterials for chemical sensing, *Advanced Materials*, 7, 1995, 375–378.

45. F. Gong, H. Wang, X. Xu, G. Zhou and Z.-S. Wang, In situ growth of $Co_{0.85}Se$ and $Ni_{0.85}Se$ on conductive substrates as high-performance counter electrodes for dye-sensitized solar cells, *Journal of American Chemical Society*, 134, 2012, 10953–10958.

46. V. Colvin, M. Schlamp and A. Alivisatos, Light-emitting diodes made from cadmium selenide nanocrystals and a semiconducting polymer, *Nature*, 370, 1994, 354–357.

47. V. Nicolosi, M. Chhowalla, M. G. Kanatzidis, M. S. Strano and J. N. Coleman, Liquid exfoliation of layered materials, *Science*, 340, 2013, 1–18.

48. Q. H. Wang, K. Kalantar-Zadeh, A. Kis, J. N. Coleman and M. S. Strano, Electronics and optoelectronics of two-dimensional transition metal dichalcogenides, *Nature Nanotechnology*, 7, 2012, 699–712.

49. L. T. L. Lee, J. He, B. Wang, Y. Ma, K. Y. Wong, Q. Li, X. Xiao and T. Chen, Few-layer $MoSe_2$ possessing high catalytic activity towards iodide/tri-iodide redox shuttles, *Scientific Reports*, 4, 2014, 4063.

50. H. Chen, Y. Xie, H. Cui, W. Zhao, X. Zhu, Y. Wang, X. Lu and F. Huang, In situ growth of a $MoSe_2$/Mo counter electrode for high efficiency dye-sensitized solar cells, *Chemical Communications*, 50, 2014, 4475–4477.

51. X. Yuan, B. Zhou, X. Zhang and L. Liu, Hierarchical $MoSe_2$ nanoflowers used as highly efficient electrode for dye-sensitized solar cells, *Electrochimica Acta*, 283, 2018, 1163–1169.

52. J. Guo, S. Liang, Y. Shi, C. Hao, X. Wang and T. Ma, Transition metal selenides as efficient counter electrode materials for dye-sensitized solar cells, *Physical Chemistry Chemical Physics*, 17, 2015, 28985–28992.

53. W. Zhu, J. Y. Yang, X. H. Gao, S. Q. Bao, X. A. Fan, T. J. Zhang and K. Cui, Effect of potential on bismuth telluride thin film growth by electrochemical atomic layer epitaxy, *Electrochimica Acta*, 50, 2005, 4041–4047.

54. S. Zhou, Q. Jiang, J. Yang, W. Chu, W. Li, X. Li, Y. Hou and J. Hou, Regulation of microstructure and composition of cobalt selenide counter electrode by electrochemical atomic layer deposition for high performance dye-sensitized solar cells, *Electrochimica Acta*, 220, 2016, 169–175.

55. C. Bao, F. Li, J. Wang, P. Sun, N. Huang, Y. Sun, L. Fang, L. Wang and X. Sun, One-pot solvothermal in situ growth of 1D single-crystalline NiSe on Ni foil as efficient and stable transparent conductive oxide free counter electrodes for dye-sensitized solar cells, *ACS Applied Materials & Interfaces*, 8, 2016, 32788–32796.

56. L. Chen, H. Yin, Y. Zhou, H. Dai, T. Yu, J. Liu and Z. Zou, In situ direct growth of single crystalline metal (Co, Ni) selenium nanosheets on metal fibers as counter electrodes toward low-cost, high-performance fiber-shaped dye-sensitized solar cells, *Nanoscale*, 8, 2016, 2304–2308.

57. C.-T. Li, C.-P. Lee, I.-T. Chiu, R. Vittal, Y.-J. Huang, T.-Y. Chen, H.-W. Pang, J. T. Lin and K.-C. Ho, Hierarchical $TiO_{1.1}Se_{0.9}$-wrapped carbon cloth as the TCO-free and Pt-free counter electrode for iodide-based and cobalt-based dye-sensitized solar cells, *Journal of Materials Chemistry A*, 5, 2017, 14079–14091.

58. Y. Zhao, J. Duan, Y. Duan, H. Yun and Q. Tang, 9.07%-efficiency dye-sensitized solar cell from Pt-free RuCoSe ternary alloy counter electrode, *Materials Letters*, 2018, 2018, 76–79.

59. J. Theerthagiri, R. A. Senthil, M. H. Buraidah, M. Raghavender, J. Madhavan and A. K. Arof, Synthesis and characterization of $(Ni_{1-x}Co_x)Se_2$ based ternary selenides as electrocatalysts for triiodide reduction in dye-sensitized solar cells, *Journal of Solid State Electrochemistry*, 238, 2016, 113–120.

60. Y.-F. Du, J.-Q. Fan, W.-H. Zhou, Z.-J. Zhou, J. Jiao and S.-X. Wu, One-step synthesis of stoichiometric $Cu_2ZnSnSe_4$ as counter electrode for dye-sensitized solar cells, *ACS Applied Materials & Interfaces*, 4, 2014, 1796–1802.

61. S.-J. Yuan, Z.-J. Zhou, Z.-L. Hou, W.-H. Zhou, R.-Y. Yao, Y. Zhao and S.-X. Wu, Enhanced performance of dye-sensitized solar cells using solution-based in situ synthesis and fabrication of $Cu_2ZnSnSe_4$ nanocrystal counter electrode, *Chemistry – A European Journal*, 19, 2013, 10107–10110.

62. F. Ozel, A. Sanlmaz, B. Istanbullu, A. Aljabour, M. Kus and S. Sonmezoglu, Penternary chalcogenides nanocrystals as catalytic materials for efficient counter electrodes in dye-sensitized solar cells, *Scientific Reports*, 6, 2016, 29207.

63. L. Zhou, K. Xu, A. Zubair, A. D. Liao, W. Fang, F. Ouyang, Y.-H. Lee, K. Ueno, R. Saito, T. Palacios, J. Kong and M. S. Dresselhaus, Large-area synthesis of high-quality uniform few-layer $MoTe_2$, *Journal of the American Chemical Society*, 137, 2015, 11892–11895.
64. E. Zhang, R. Chen, C. Huang, J. Yu, K. Zhang, W. Wang, S. Liu, J. Ling, X. Wan, H.-Z. Lu and F. Xiu, Tunable positive to negative magnetoresistance in atomically thin WTe_2, *Nano Letters* 17, 2017, 878–885.
65. C. Ruppert, O. B. Aslan and T. F. Heinz, Optical properties and band gap of single- and few-layer $MoTe_2$ crystals, *Nano Letters*, 14, 2014, 6231–6236.
66. T.-R. Chang, S.-Y. Xu, G. Chang, C.-C. Lee, S.-M. Huang, B. Wang, G. Bian, H. Zheng, D. S. Sanchez, I. Belopolski, N. Alidous, M. Neupane, A. Bansil, H.-T. Jeng, H. Lin and M. Zahid Hasan, Prediction of an arc-tunable Weyl Fermion metallic state in $Mo_xW_{1-x}Te_2$, *Nature Communications*, 7, 2016, 10639.
67. S. Hussain, S. A. Patil, D. Vikraman, N. Mengal, H. Liu, W. Song, K.-S. An, S. H. Jeong, H.-S. Kim and J. Jung, Large area growth of $MoTe_2$ films as high performance counter electrodes for dye-sensitized solar cells, *Scientific Reports*, 8, 2018, 29.
68. H.-W. Tsai, T.-H. Wang, T.-C. Chan, P.-J. Chen, C.-C. Chung, A. Yaghoubi, C.-N. Liao, E. W.-G. Diau and Y.-Lun Chueh, Fabrication of large-scale single-crystal bismuth telluride (Bi_2Te_3) nanosheet arrays by a single-step electrolysis process, *Nanoscale*, 6, 2014, 7780–7785.
69. R. J. Mehta, Y. Zhang, C. Karthik, B. Singh, R. W. Siegel, T. Borca-Tasciuc and G. Ramanath, A new class of doped nanobulk high-figure-of-merit thermoelectric by scalable bottom-up assembly, *Nature Materials*, 11, 2012, 233–240.
70. T. Chen, G. H. Guai, C. Gong, W. Hu, J. Zhu, H. Yang, Q. Yan and C. M. Li, Thermoelectric Bi_2Te_3 improved charge collection for high-performance dye-sensitized solar cells, *Energy & Environmental Science*, 5, 2012, 6294–6298.
71. J. Jia, J. Wu, J. Dong and J. Lin, Cobalt telluride/reduced graphene oxide using as high performance counter electrode for dye-sensitized solar cells, *Electrochimica Acta*, 185, 2015, 184–189.

Chapter 7

Photovoltaics Performance of Carbon Nanotubes and Their Composites Based Dye-Sensitized Solar Cells

A. Dennyson Savariraj[a] and R. V. Mangalaraja[b]
[a]*Advanced Ceramics and Nanotechnology Laboratory, Department of Materials Engineering, Faculty of Engineering, University of Concepcion, Concepcion 4070409, Chile*
[b]*Technological Development Unit (UDT), University of Concepcion, Coronel Industrial Park, Coronel 4191996, Chile*
dennysons@gmail.com

Dye-sensitized solar cells (DSSCs) emerged as a cost-effective alternative to conventional silicon-based solar cells, which made DSSC to be the next generation of photovoltaic technologies and research hotspots with a promising future. The different components that make DSSC are photoanode, consisting of wide bandgap materials and dye, an electrolyte, and a counter electrode (CE). Among them, the counter electrode is a critical component that collects electrons from the external circuit and catalyzes the redox reduction in the iodine iodide electrolyte. So, it plays a crucial role in determining the photovoltaic performance, cost of the device, and

Counter Electrode for Dye-Sensitized Solar Cells
Edited by Alagarsamy Pandikumar, Ramesh Mohan, and Kandasamy Jothivnekatachalam

ISBN 978-981-4877-38-1 (Hardcover), 978-1-003-11077-4 (eBook)
www.jennystanford.com

its stability. Usually, platinum (Pt) is used as a CE to regenerate the dye by regenerating I^- from I_3^- (iodine iodide redox couple) species in the redox couple. Pt being an expensive noble metal; there is a need to find cost-effective alternative materials to replace Pt as the CE. This chapter presents a summary of carbon nanotube (CNT)-based composites to be used as a CE for DSSC.

7.1 Introduction

DSSC, a third-generation solar cell, is facing several challenges, and one among them is the cost factor due to the use of Pt as CE material. Over the last two decades, there have been several attempts made to replace CE material to improve photovoltaic performance [1, 2]. Platinum has been widely used as the CE material in DSSC as it can function as a catalyst to generate I^- from I_3^- and bring electrons to the electrolyte from the external load. The exorbitant cost of Pt and its scarcity make it inevitable to look for an alternative material to replace Pt as CE material. Moreover, producing the best Pt CE for DSSC requires a high-temperature hydrolysis process, which is a handicap to deposit on a flexible substrate [3]. The alternate materials include polymers, transition metal chalcogenides, and carbides, polymers, carbon black, graphene, and CNTs [4]. Among several alternative materials, carbon-based materials are the most suitable replacement of Pt due to their enhanced reduction capability with iodine/iodide electrolyte, chemical stability, high electrical conductivity, and surface area [5, 6]. The different carbon materials used as CE are carbon black [7], activated carbon [8], graphite [9], hard carbon spherules [10], and CNTs [11–14]. CNTs got their name from their hollow and long structure whose walls are made up of one-atom-thick sheets of carbon. Further, carbon is one of the most abundant elements in the earth's crust, and it can be a suitable replacement for Pt as a CE in DSSC. Its abundance as the sixth material in the earth's crust makes it readily available and economical. Besides being cheap, the apparent properties of carbon such as high catalytic activity on the redox species, resistance to corrosion, environmental friendliness make carbon materials a suitable alternative to Pt. The different types of carbon materials used as counter electrode are carbon black, mesoporous carbon, activated carbon, graphene, fullerenes, and CNTs [4]. Despite having less electrocatalytic activity than Pt, carbon-based materials may still be the best alternative to Pt because they have high porosity and large surface area. The very first attempt of using carbon materials was made by Gratzel et al. in 1996, where they used a mixture of carbon black and graphite, which replaced Pt as a counter electrode yielding the highest photoconversion efficiency of 6.7% [15].

7.2 Carbon Nanotubes

The different allotropes of carbon are diamond, graphite, and fullerene (C60, C70, etc.) [16]. The properties of different allotropes of carbon are listed in Table 7.1. CNTs are tubular form of carbon-based nanostructures that can be visualized as graphene rolled up into a seamless cylinder at the nanometer scale [17]. They are materials with a high aspect ratio of more than 1000 and the CNTs have capped one end covered with fullerene structure at the minimum [16].

The CNTs are classified into two types based on the fabrication process, i.e., single-walled CNTs (SWCNTs) and multi-walled CNTs (MWCNTs) [18–21]. The main difference between SWCNTs and MWCNTs is the number of shells. SWCNTs are made up of single graphene layer sheet rolled up into a seamless cylinder. On the contrary, MWCNTs are made up of more than one graphene sheet rolled up in concentric cylindrical shells, arranged coaxially around a central hollow core and the van der Waals forces act between the adjacent layers [16].

CNTs have many advantages which include the availability of raw materials in abundance (carbon), chemical stability, and significant inherent electrical properties at room temperature with high carrier mobility and conductivity. Further, they possess a broad spectral range of transmittance with a neutral colour and mechanical flexibility [17, 22–24].

In this chapter we are presenting the different routes of synthesis of CNTs, their structural and physical properties like mechanical, thermal, and electrical properties, and chemical properties, as well as the various processes of making CNT/polymer composites such as in situ polymerization, melt processing, and solution mixing [25].

Table 7.1 Physical properties of different carbon materials [16]

	C Material				
Property	**Graphite**	**Diamond**	**Fullerene**	**SWCNT**	**MWCNT**
Specific gravity (g/cm^3)	1.9–2.3	3.5	1.7	1.7	1.8
Electrical conductivity (S/cm)	4000^p, 3.3^c	10^{-2}–10^{-15}	10^{-5}	10^2–10^6	10^3–10^5
Electron mobility [$cm^2/(V\ s)$]	2.0×10^4	1800	0.5–6	$\sim 10^5$	10^4–10^5
Thermal conductivity [W/(m K)]	298^p, 2.2^c	900–2320	0.4	6000	2000
Coefficient of thermal expansion (K^{-1})	-1×10^{-6p} 2.9×10^{-5c}	$(1\sim3) \times 10^{-6}$	6.2×10^{-5}	Negligible	Negligible
Thermal stability in air (°C)	450–650	<600	~600	>600	>600

p: in-plane; c: c-axis

The CNTs are grown to a length of few micrometres, and their diameters go up to 100 nm. The CNTs formed, usually lead to the formation of bundles and get entangled, making a complex network structure. The conductivity of CNTs can be either metallic or semiconducting depends on the arrangement of the hexagon rings along with the tubular furnace. The usage of CNTs is curtailed because of their lack of solubility and constrains for manipulation. The pristine CNTs are insoluble in both aqueous and organic solvents; however, they can only be dispersed in selective solvents using mechanical forces such as continuous sonication which would otherwise end up with precipitation. CNTs can be used for most applications only after functionalization by subjecting them to chemical reactions after which they are soluble. The functionalization of the CNTs are classified into three categories:

(a) the covalent attachment of chemical groups through reactions onto the ð-conjugated skeleton of CNT, (b) the noncovalent adsorption or wrapping of various functional molecules, and (c) the endohedral filling of their empty inner cavity [26].

7.3 Synthesis of CNTs

The CNTs to be used for fundamental and technological applications need to be of high-quality which implies that the CNTs must be free from structural and chemical defects along the tube axes over a significant length that might be up to 10 μm. Any synthesis method is successful only if it can bring out cost-effective mass production of high-quality nanotubes and do selective output with control over obtaining CNTs of desired properties. CNTs are mostly produced through different methods which include arc evaporation, chemical vapour deposition (CVD), laser ablation, flame synthesis, electrolysis etc. The three techniques mostly adopted for the synthesis of CNTS are arc deposition, laser ablation, and chemical vapour deposition (CVD). The essential materials for the production of CNTs are a source of carbon, catalyst, and required energy. The energy source will generate electricity for the arc discharge method, heat for CVD, and high-density light for laser ablation [27].

7.3.1 Arc Discharge

The arc discharge is the electrical breakdown of gas to generate plasma, and Iijima used this method for the preparation of CNTs [28]. The main requirement is the condensation of hot gaseous carbon atoms produced by the evaporation of solid carbon [29]. The basic principle behind this method is, creating a hot plasma discharge between two electrodes connected to the power supply (100 A; 20 V) in the presence of Helium (He) gas to produce CNTs. High yield of CNTs is achieved by increasing the pressure of He to a certain level and increasing it further brings down the yield [30]. The quality of CNTs produced can be improved by lowering the current [31]. Further, replacing He gas by N_2 [32] or CF_4 [33] and using organic vapours can increase the yield [34].

A schematic diagram of an arc discharge chamber for the synthesis of CNTs is shown in Fig. 7.1. The reaction chamber has two electrodes namely anode which is filled with carbon powder precursor with catalyst and cathode is made up of pure graphite rod. These electrodes are positioned either vertically or horizontally. The entire chamber is either submerged inside a liquid environment or filled with gas. The electrodes are brought together and kept in close proximity of 1–2 mm after giving the power supply (AC or DC). The gap between the two electrodes is maintained using closed-loop automation to keep a non-fluctuating arc with a constant current which would otherwise affect both the quantity and quality of the product. The carbon precursor at the anode in the reaction chamber sublimes due to the plasma generated with a high temperature ~4000–6000 K. The carbon vapour generated aggregate in the gas phase and moves to the cathode, and the temperature gradient cools down at the cathode. The deposited CNTs are collected from the walls of the chamber. The reaction and the yield are affected by different parameters such as temperature, voltage, current, frequency, pressure, shape, cathode type, etc. [28].

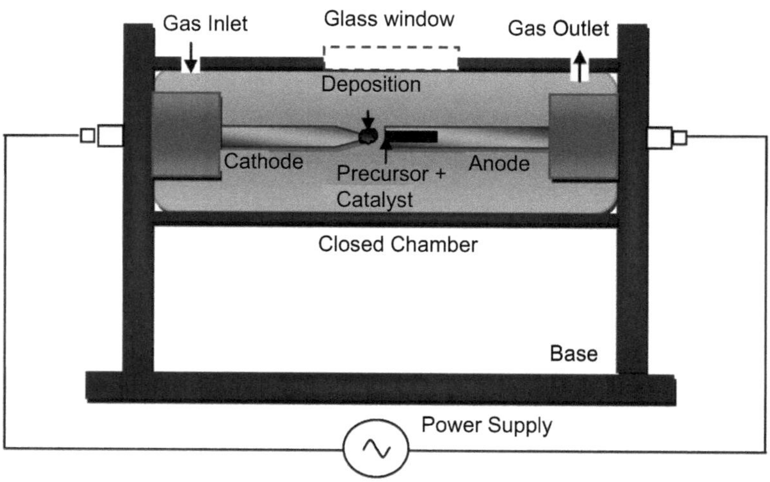

Figure 7.1 Schematic illustration of an arc discharge setup for the production of CNTs [28].

7.3.2 Laser Ablation Method

Ever since Richard Smalley reported the synthesis of SWCNTs using laser ablation method in 1995, this method gained momentum. The basic working principle and mechanisms are very similar to the arc discharge method excepting the laser energy source is replaced with electricity. This method makes use of graphite as the carbon source which is mixed with metal catalysts such as Co and Ni in small quantity and subjected to vaporization using a pulsed laser beam in the presence of inert gas at a high temperature of 1200°C using a constant pressure of 500 Torr [35]. The laser can be operated both in continuous mode or pulsed mode to vaporize carbon along with catalyst and collected at the water-cooled copper to which a flow of inert gas carries them. To obtain a reasonable yield, certain parameters have to be tuned as follows. The wavelength should be adjusted so that it has minimum absorption depth that can focus the total energy in a small volume resulting in rapid and complete ablation. To minimize the conduction loss and to maximize peak power, there should be short pulse duration with high repetition rate and with high beam quality which includes focusability, brightness, and homogeneity [36]. Some researchers achieve an increase in the yield by incorporating double pulsed laser to boost vaporization [37]. However, the requirement of costly lasers and high-power supply make this method very expensive [28]. A schematic representation of CNT production by the laser ablation method is given in Fig. 7.2.

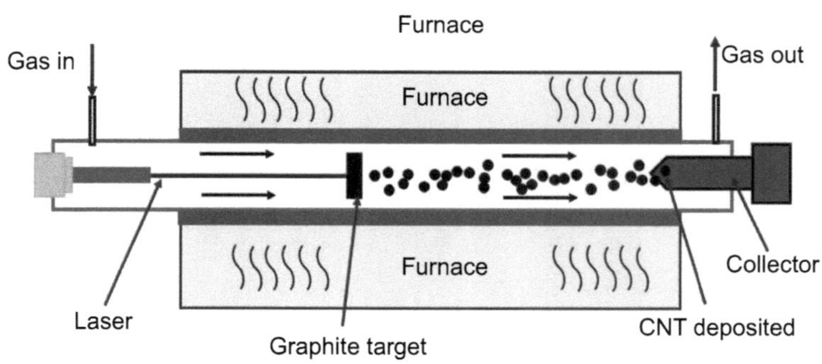

Figure 7.2 Schematic illustration for carbon nanotube synthesis by the laser ablation method.

Table 7.2 Comparison of prime parameters for the three fabrication methods [36]

Method	Operating temperature (°C)	Length of CNT (μm)	Growth rate of CNT (μms^{-1})	Yield	Quality	Purity
Arc discharge	~4000	~1	up to 10^7	Low	Low	Low
Laser ablation	Room temperature to 1000	~1	~0.1	Low	High	Medium
Thermal CVD	500–1200	0.1 to 10^5	0.1–10	High	Medium	Medium to Low
PECVD	100–800	0.1–10	0.01–1	Low	Low-Medium	Medium

Table 7.3 A summary of the three primary production methods and their efficiency [27]

Method	Arc discharge	Laser ablation	Chemical vapour deposition
Process	Connect two graphite rods to a power supply, place them a few millimetres apart. At 100 A, carbon vaporizes and forms hot plasma.	Blast graphite with intense laser pulses; use the laser pulses rather than electricity to generate carbon gas from which the CNTs form; try various conditions until hit on one that produces prodigious amounts of SWNTs.	Place substrate in oven, heat to high temperature, and slowly add a carbon-bearing gas such as methane. As gas decomposes it frees up carbon atoms, which recombine in the form of NTs.

Method	Arc discharge	Laser ablation	Chemical vapour deposition
Condition	Low-pressure inert gas (Helium)	Argon or Nitrogen gas at 500 Torr	High temperatures within 500 to 1000°C at atmospheric pressure
Typical yield	30–90%	Up to 70%	20–100%
SWCNT	Short tubes with diameters of 0.6–1.4 nm.	Long bundles of tubes (5–20 microns), with individual diameter from 1–2 nm.	Long tubes with diameters ranging from 0.6 to 4 nm.
MWCNT	Short tubes with inner diameter of 1–3 nm and outer diameter of approximately 10 nm.	Though it is possible to synthesize MWNT, it is less preferred since it is expensive.	Long tubes with diameter ranging from 10 to 240 nm.
Carbon source	Pure graphite	Graphite	Fossil-based hydrocarbon and botanical hydrocarbon
Cost	High	High	Low
Advantage	Can easily produce SWNT, MWNTs. SWNTs have few structural defects; MWNTs without catalyst, not too expensive, open air synthesis possible.	Good quality, higher yield, and narrower distribution of SWNT than arc-discharge.	Easiest to scale up to industrial production; long length, simple process, SWNT diameter controllable, and quite pure.
Disadvantage	Tubes tend to be short with random sizes and directions; often needs a lot of purification.	Costly technique, because it requires expensive lasers and high-power requirement, but is improving.	Often riddled with defects.

7.3.3 Chemical Vapour Deposition (CVD)

CVD is more preferred over arch discharge and laser ablation method because the in situ synthesis of CNTs can be made at a lower temperature. Figure 7.3 presents the schematic representation for the production of CNTs by the CVD method. This method is most suited over the previously discussed methods for the controlled growth and bulk production of CNTs [29]. The fundamental principle of CVD is the catalytic assisted decomposition of a mixture of hydrocarbon gases (methane, ethylene, etc.) or volatile carbon compounds and the catalysts are usually transition metals such as Fe, Co, Ni, and their alloys taken in the form of film. The metal nanoparticles play a dual role of catalyst and nucleation site. The type and the quality of the CNTs are determined by the kind of catalyst and preparation [38, 39]. This process is typically operated at a lower temperature between 700°C and 900°C and depending upon the nature of carbon source and the catalyst. The catalysis process for CNT can be either homogeneous or heterogeneous. There are reports on the synthesis of CNTs using organometallic compounds instead of hydrocarbon gas [40], and it has the advantage of skipping the removal of catalyst step as the carbon source, and the catalyst is in the same phase. The ideal substrate for the controlled growth of CNTs is porous silicon. The modified form of CVD used for the production of CNTs is plasma-enhanced chemical vapour deposition (PECVD) [41], which makes use of a DC plasma to align CNTs for flat panel displays and solar cells [42, 43]. During the CVD process, the different parameter of CNTs structure from wall number to length (up to centimetres), alignment and diameter can be controlled [44]. Therefore, CVD has the advantage of low cost, controllable process and mild oxidation, however the CNTs produced do have defects and their crystalline structure can be recovered at temperature near to 2000°C [45].

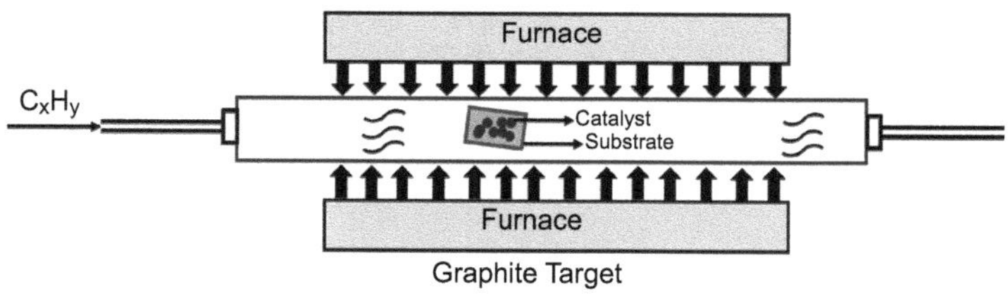

Figure 7.3 Schematic diagram of chemical vapour deposition setup for the production of CNTs.

A detailed report of the different parameters for the three primary fabrication methods is given in Table 7.2 and Table 7.3, along with the details of the three methods, their advantages, disadvantages, and efficiencies.

7.4 Properties of CNTs

The diameters of SWCNTs and MWCNTs are 0.8–2nm and 5–20 nm respectively, and their length can vary from less than 100 nm to few centimetres. The nanotubes are constructed with the length-to-diameter ratio of up to 132,000,000:1 owing to their exceptional strength and stiffness, which is much better than several other materials [46, 47]. The fundamental properties of CNTs are presented in Table 7.4, and the differences between SWCNT and MWCNT are given in Table 7.5.

Table 7.4 Summary of Properties of CNTs [48]

Property	Value	Remarks
Intrinsic mobility	Exceeding 10^5 cm^2 V^{-1} s^{-1} for individual CNTs at room temperature [49, 50]	Almost 100 times higher than that of silicon at 300 K with dopant concentration at 10^{17} cm^{-3} [51–53]
Free carriers concentration	~10^{17} cm^{-3} [54, 55]	Lower than that of graphene (~10^{20} cm^{-3}), TCOs (~10^{21} cm^{-3}), and most of metals (such as ~10^{22} cm^{-3} for silver) [56, 57]
Current carrying capacity	Exceeding 10^9 A cm^{-2} for individual CNT [58]	1000 times higher than copper [59]
Electrical conductivity	10^4–10^6 S cm^{-1} [60–63]	Close to that of some metals, such as mercury [64]
On/off current ratios	Higher than 10^5 [65–67]	1000 times higher than a bilayer structure at room temperature [68]
Thermal conductivity	Up to 3500 W m^{-1} K^{-1} [69–71]	About 1500 W m^{-1} K^{-1} higher than that of diamond [72, 73]
Young's modulus	1–2 TPa [74–80]	Comparable to that of a single crystal diamond at room temperature [81]
Fracture stress	50 GPa [82–84]	~50 times larger than that of steel wires after density normalization [85, 86]
Surface area	1600 m^2 g^{-1} [87, 88]	Higher than that of activated carbon (1200 m^2 g^{-1}) [89, 90]

Table 7.5 Comparison between SWCNT and MWCNT [27]

SWCNT	MWCNT
Single-layer of graphene.	Multiple layers of graphene.
Catalyst is required for synthesis.	Can be produced without the catalyst.
Bulk synthesis is difficult as it requires proper control over growth and atmospheric condition.	Bulk synthesis is easy.
Bulk synthesis is difficult as it requires proper control overgrowth and atmospheric condition.	Bulk synthesis is facile.
Resistivity usually in the range of 10^{-4}–10^{-3} Ω·m.	Resistivity usually in the range of 1.8×10^{-5}–6.1×10^{-5} Ω·m.
Purity is poor. Typical SWCNT content in as-prepared samples by chemical vapour deposition (CVD) method is about 30–50 wt%. However, high purity up to 80% have been reported by using arc discharge synthesis method.	Purity is high. Typical MWCNT content in as-prepared samples by the CVD method is about 35–90 wt%.
A chance of defect is more during functionalization.	A chance of defect is less, mainly when synthesized by the arc-discharged method.
Characterization and evaluation are easy.	It has a very complex structure.
It can be easily twisted and is more pliable.	It cannot be easily twisted.

7.5 Carbon Nanotubes/Polymer Nanocomposites

CNT/polymer composites have the advantage of combining the properties of both the components and therefore, they do have a high demand for several applications including solar cells. The CNTs are usually embedded or incorporated into the polymer matrix as filler materials, and it can be done in the following ways.

7.5.1 Preparation of Carbon Nanotubes/Polymer Nanocomposites

In the CNT/polymer composite, a polymer is the matrix and CNT is the filler. Developing composite materials requires a homogeneous mixture between the filler and the matrix along with strong interaction and adhesion between them. The different processes for the development of CNT/polymer composites are: (1) solution mixing, (2) melt processing, and (3) in situ polymerization.

7.5.1.1 Solution mixing

Solution mixing is one of the simplest methods to make CNT/polymer nanocomposite. It involves the suspension of CNTs and polymer in a suitable solvent to form a nanocomposite film on the surface of the substrate, followed by evaporating the solvent in a controlled atmosphere. Some of the polymers processed by this method are PMMA, PVA, and polystyrene [91–93]. The CNT/polymer composite can also be prepared better by including different variations like employing an ultra-sonication method for dispersing MWNTs with polystyrene [93, 94]. In some cases, surface modification of the CNTs by different functional groups facilitates proper dispersion; however, at times they lead to compatibility problem of the functional group within the polymer matrix that can be addressed by addition of surfactants to keep CNTs intact [95–97]. Sometimes the problem of agglomeration after evaporation of the solvent is a severe issue; spin coating or electrospinning the composite suspension can address this issue [91, 98, 99].

7.5.1.2 Melt processing

The main drawback of the solution mixing method is the limitation of the solubility of the polymers in the solvent. Therefore, melt mixing method is used to overcome this shortfall which uses thermoplastic polymers, e.g., polycarbonate, polystyrene, polypropylene, poly (ethylene-2,6-naphthalate), etc. [100–102]. The thermoplastic polymers are melted to a viscous liquid and then blended with CNTs. Extrusion and injection modelling techniques are used to make shear mixing for improving the dispersion of CNTs [103]. Although this technique is advantageous over solution mixing in terms of improved solubility, uniform dispersion of CNTs is a challenge due to the high viscosity of thermoplastic polymers [104].

7.5.1.3 In situ polymerization

In situ polymerization involves dispersion of the CNTs into the monomer matrix either with or without solvent and polymerization is carried out. This technique is most suited for polymers which are insoluble and thermally unstable and cannot be processed both by solution mixing and melt processing. This technique makes use of monomers as starting material instead of polymers. Therefore, it has the advantage of yielding better dispersion coefficients and better interaction between the host (polymeric matrix) and the guest (CNTs) by grafting of the polymer molecules on to the CNTs. Some reported CNT/polymer composites using this method are WNT/polystyrene, MWCNT/polyurethane, MWCNT/polyethylene, MWCNT/polypyrrole, MWCNT/nylon, etc. [25, 105–108].

7.6 Carbon Nanotubes–Polymer Composites as Counter Electrodes for DSSC

7.6.1 CNT-Based Counter Electrodes

Fabrication of composite electrodes often leads to a combination of properties of all of its constituents. The composite electrodes have higher catalytic activity, increased stability towards corrosion, high charge transfer, reduced impedance, increased surface area, etc., which enhances the overall PCE (power conversion efficiency) of the cell and its longevity. Pure CNTs as such can also function as CE. Suzuki et al. in 2003 explored the very idea of incorporating commercially available SWCNTs as CE for DSSCs, where they reported a high PCE of 4.5%, which is 85% of the PCE exhibited by DSSCs with Pt-sputtered CE [11]. Choi et al. in 2010, attempted in incorporating MWCNTs as CE along with its composite with poly(maleic acid-co-phydroxystyrene)-block-poly(p-hydroxystyrene) (HSPM) [109]. Table 7.6 summarises the different solar cell parameters obtained for various CE made of CNT, CNT/polymer composite, and Pt.

Table 7.6 Different solar cell parameters for DSSC fabricated using CNT-, CNT/polymer composite-, and Pt-based counter electrodes

Material	Active area (cm^2)	Dye	DSSC with CNT/Polymer				DSSC with Pt				Ref.
			V_{OC} (V)	J_{SC} (mA/cm^2)	FF	η (%)	V_{OC} (V)	J_{SC} (mA/cm^2)	FF	H (%)	
MWCNT	-	N719	0.764	12.93	0.57	5.69	0.760	12.96	0.72	7.15	[109]
MWCNT	0.36	N719	0.720	8.41	0.71	7.96	0.710	8.60	0.71	8.03	[13]
SWCNT	-	-	0.680	13.70	0.26	2.34	-	-	-	-	[110]
SWCNT/PVP	-	-	0.718	12.14	0.51	4.50	-	-	-	-	[110]
SWCNT	-	N719	0.742	14.25	0.74	7.81	0.746	14.63	0.73	7.92	[12]
SWCNT/CMC	-	N719	0.738	13.92	0.66	6.81	0.746	14.63	0.73	7.92	[12]
MWCNT	-	N719	0.747	14.49	0.71	7.63	0.746	14.63	0.73	7.92	[12]
MWCNT/CMC	-	N719	0.746	14.54	0.66	7.17	0.746	14.63	0.73	7.92	[12]
MWCNT	-	N719	0.74	15.54	0.57	6.62	0.79	16.27	0.68	8.80	[111]
Paste printed MWCNT	0.25	N719	0.738	15.27	0.69	8.03	0.746	17.68	0.65	8.80	[112]
CVD grown MWCNT	0.25	N719	0.755	17.62	0.73	10.04	0.746	17.68	0.65	8.80	[112]

(*Continued*)

Table 7.6 (*Continued*)

Material	Active area (cm^2)	Dye	DSSC with CNT/Polymer				DSSC with Pt				Ref.
			V_{OC} (V)	J_{SC} (mA/cm^2)	FF	η (%)	V_{OC} (V)	J_{SC} (mA/cm^2)	FF	H (%)	
MWCNT	-	N719	0.764	12.93	0.57	5.69	0.760	12.96	0.72	7.15	[109]
MWCNT/HSPM	-	N719	0.753	12.62	0.62	5.94	0.760	12.96	0.72	7.15	[109]
MWCNTs/PEDOT	0.196	N719	0.74	11.5	0.64	5.47	0.76	11.3	0.63	5.43	[113]
PVC/SWCNTs/PEDOT (FTO-free)	0.54	-	0.70	15.3	0.47	5.00	0.61	16.0	0.55	5.2	[114]
MWCNT/PEDOT:PSS	0.25	N719	0.66	15.5	0.63	6.5	0.74	17.9	0.64	8.5	[3]
MWCNT/PSSA	-	-	0.60	9.7	0.62	3.6	0.74	17.9	0.64	8.5	[3]
MWCNTs/PSSNa	-	N719	0.71	16.9	0.59	7.03	0.72	17.7	0.65	8.3	[115]
MWCNTs/PTh	0.15	N719	0.70	10.5	0.58	4.72	0.75	10.8	0.63	5.68	[116]
MWCNTs/ Polypropylene	0.12	N719	0.82	13.8	0.59	6.67	0.78	14.0	0.71	7.72	[117]
MWCNTs/PANI	0.25	N719	0.78	16.1	0.58	7.21	0.77	17.2	0.58	7.59	[118]
MWCNTs/polypyrrole	0.28	N719	0.77	14.8	0.65	7.42	0.76	14.5	0.62	6.85	[119]

The pristine MWCNT yielded a PCE of 5.69% while MWCNT/HSPM composite yielded a slightly higher PCE of 5.94% against 7.15% from Pt CE [109]. Velten et al. used the dry spun carbon MWCNT sheets on FTO employed it as CE for DSSC and reported a high PCE of 6.62% Vs 8.80% for Pt used as reference [120]. Lee et al. obtained a PCE for MWCNT as CE with 7.96% while 8.03% for Pt as CE in DSSC. They attributed the increased number of defect-rich planes in bamboo-like structured MWCNT as the reason behind increased PCE, fill factor (FF) values, and lower charge transfer resistance. The lowering of charge transfer resistance led to the highest PCE for MWCNT as CE and lies very close to that of the Pt [13].

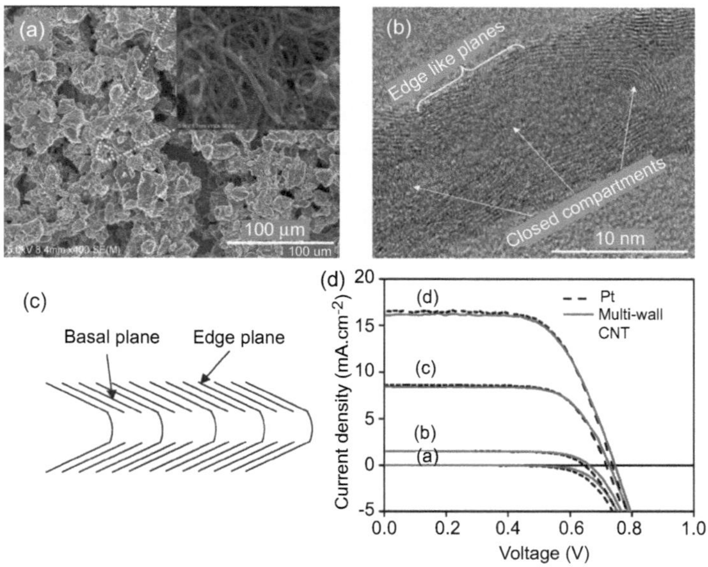

Figure 7.4 Microstructural characterization of multi-wall CNTs: (a) FE-SEM image of a multi-wall CNT/CMC composite film on a fluorine-doped SnO_2-conducting glass substrate. Inset: enlarged view of the marked portion. (b) TEM image of the bamboo-like structure in multi-wall CNTs used in this study. (c) Schematic of the bamboo-like structure multi-wall CNT. (d) Current-voltage characteristics of DSSC with multi-wall CNT (solid line) and platinum (dashed line) counter electrodes. Curve (a) shows the performance of devices in the dark. Curves (b)–(d) were measured under 0.1, 0.5, and 1 sun illumination (air mass 1.5 G), respectively [13].

Figure 7.4 shows the MWCNTs aggregations of different size distributions on FTO due to the high concentration of MWCNTs.

The aggregates of MWCNT mostly consist of randomly oriented mesoporous nanotube work, as shown in the inset in Fig. 7.4a. From the TEM images shown in Fig. 7.4b, it is evident that the MWCNTs possess bamboo-like structure. The edge planes at regular intervals along the tube surface arise because of the presence of a large number of transverse walls with periodically closed compartments. Figure 7.4d gives the current-voltage performance of multi-wall CNT counter electrode DSSCs in the dark and under the illumination of different levels of solar illumination.

7.6.2 Carbon Nantotube/Polymer Composite–Based Counter Electrode

Park et al. made a comparative study between SWCNT and its composite with polyvinyl pyrrolidone (PVP) as CE materials for DSSC. They added 1wt% of sodium dodecyl sulfate (SDS) to SWCNT in 100 ml of anhydrous ethanol and sonicated it for 1 h. Similarly, they added 1wt% of PVP to the SWCNT dispersed and sonicated for 2 h followed by incubation at 50°C for 12 h. Also, they used carboxyl methyl cellulose (CMC) as a binder and stirred for 16 h to get a paste. Using a simple casting method, they deposited the SWCNT and SWCNT/PVP paste and calcined at 300°C for 1 h to get rid of PVP, CMC, and SDS. The DSSCs assembled with SWCNT as CE exhibited a PCE of 2.34%, whereas the PCE of DSSC with SWCNT/PVP as CE was almost two times of the former (4.5%). They attributed the high PCE by SWCNT/PVP-based CE due to higher electrocatalytic activity arising from the uniform and smooth surface of SWCNT on adding PVP. Here PVP seemed to have played a role as wrapping agent, sustaining the original structure of SWCNT avoiding aggregation and breakage of SWCNT. The role of PVP as wrapper and calcination both play a key role in determining the surface of the CE in such way to have a higher electrochemical reaction for charge transfer reactions and reduced resistance when compared to bare CE consisting of only SWCNT [110].

Xiaoguang et al. made a comparison using the SWCNTs and MWCNTs as CE materials in DSSC. They mixed poly(ethylene glycol) (PEG), with SWCNTs and MWCNTs and ground in an agate mortar for 10 min followed by sonicating it for 10 min. The obtained mixture was further centrifuged to remove the excess PEG. The obtained paste was used to fabricate counter electrode by doctor blade method and heated to 250°C for 10 min to remove PEG and sintered at 450°C for 30 min in air. The surface of these binder-free CNT films was rough and adhered well to the substrate. A set of electrodes were also made of only CNTs for comparison using (carboxymethyl)-cellulose (CMC) sodium salt as a binder. The DSSCs fabricated using CEs made of binder-free SWCNTs, and MWCNT gave higher PCE than the binder-based ones because the binder-based CEs possess high series resistance. This increase in the resistance is due to the insulating binder and reduction in the surface because of binder covering. The DSSC with SWCNT CEs possess low charge transfer resistance (R_{CT}) of 0.6 Ωcm^2 with a PCE of 7.81% which is higher than those devices with MWCNTs whose R_{CT} is 0.75 Ωcm^2 and PCE is 7.63 % indicating that SWCNT effectively catalyzes the redox of iodide and triiodide. The large surface area of SWCNT CEs is not the only reason for the higher PCE and low R_{CT} of the device but also is the reason for better stability when compared with MWCNT. When the efficiency of the devices was measured after four weeks, the efficiency of DSSCs increased from 7.81% to 8.17% for SWCNT devices, on the other hand, there is a decrease from 7.63% to 6.63% for MWCNT-based cells. The DSSCs with the SWCNT counter electrode exhibit better stability and enhanced photovoltaic performance than that with the conventional Pt or MWCNT CEs. The PCE of devices fabricated with binder-based SWCNT and MWCNT CEs show a lower PCE of 6.81% and 7.17% respectively, which is lower than that of those devices made of without binder [12, 47]. Velten et al. attempted to compare the electrocatalytic activity of MWCNT, graphene, and MWCNT/graphene. They reported the highest PCE for DSSCs with MWCNT/graphene composite CEs with 7.55% while the one with MWCNT CEs 6.62% which is much higher than that for graphene CEs (4.64%). The PCE is a reflection of R_{CT} values and making a composite using MWCNT/graphene seems to reduce the sheet resistance and the cell's overall internal resistance [111].

In another report by Nam et al., they explored the PCE of DSSC using MWCNT-based CEs fabricated by two different methods. MWCNTs were deposited on FTO by screen printing the randomly dispersed MWCNTs and CVD or direct growing method to make well-aligned CNTs (shown in Fig. 7.5). The DSSC with CE made of aligned MWCNTs grown by the CVD method shows a conversion efficiency of 10.04%, higher than that of DSSCs with CE with screen printed MWCNT (8.03%) and Pt itself (8.80%) under the same experimental conditions. They have attributed the higher PCE by CE from CVD-based MWCNTs to the large surface area and high electrical conductivity [47, 112].

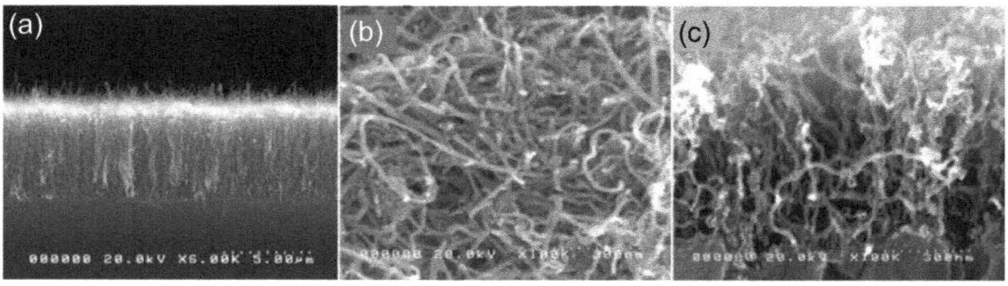

Figure 7.5 Cross-sectional SEM images of (a) CVD-grown CNT layer on a soda lime, (b) screen printing CNT paste, and (c) a directly formed CNT counter electrode [112].

Carbon nanotubes when mixed with polymers to make a composite and used as counter electrode material, show superior properties. Choi et al., reported the effect of modifying the surface of CNTs by preparing poly(p-hydroxystyrene) block copolymers (HSPM) composites by controlled/"living" radical polymerization technique and modifying the surface of MWCNT to be used as CE material for DSSSC. They prepared a stable dispersion of MWCNT/HSPM in a-terpineol and fabricated CE with uniform MWCNT coating and obtained a high PCE of 5.94% against 5.69% for DSSC with non-modified MWCNT-based CE. Here surface modification using HSPM seemed to have improved the fill factor and the PCE [109].

There are also studies made and reported successfully on the effect of using poly(3,4-ethylenedioxythiophene) (PEDOT) along with CNTs and fabricating CE with them. Zhang et al. made a composite comprising of functionalized MWCNTs and PEDOT by oxidative electropolymerization method on FTO and used it CE for DSSC. Using such CE they achieved a PCE of 5.47% for the device, which is higher than the PCE obtained for Pt-based CE (5.43%). Here through electropolymerization more porous PEDOT-MWCNT nanostructure was made so that the FTO substrate was covered both uniformly and closely with PEDOT, that increases the electrocatalytic activity and lower value of charge transfer resistance towards the I_3^-/I^- and I_2/I_3^- redox reactions than the PEDOT. The DSSC assembled with CE consisting of only PEDOT yield a PCE of 4.84% only [113]. A novel idea of transforming non-conducting polyvinyl chloride (PVC) into conducting PVC has been reported by Hashmi et al. They deposited water-based SWCNT ink on it PVC and dried it at 120°C, thus making it conductive (Fig. 7.6). The stability of the SWCNT coating and its mechanical stability was tested both by bending test and tape adhesion test. The SWCNT-coated PVC was used as a flexible CE after further modifying it by applying spin coating a thin layer of PEDOT-TsO conducting polymer over the SWCNT coating. In addition to being flexible, it is also a metal-free CE which showed a PCE of 5.0% which is almost equal to that of Pt-based CE on FTO. The charge transfer resistance (R_{CT}) for PVC/SWCNTs/PEDOT is only 2.7 Ωcm^2 as compared to 2.2 Ωcm^2 got for thermally platinized CE. The PVC/SWCNTs/PEDOT CE is not only flexible and metal-free but also proved to have almost equal electrocatalytic activity as that of Pt [114].

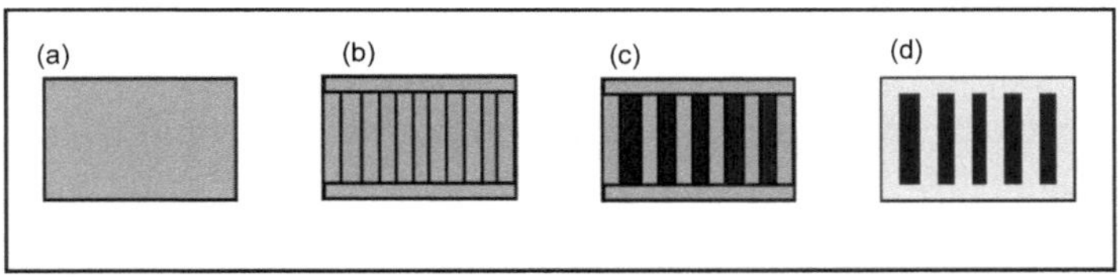

Figure 7.6 Demonstration of ink deposition: (a) patterning of the substrate, (b) application of ink and evaporation of solvent at 120°C in 5 min, and (c) SWCNT patterns.

Fan et al. carried out making a conducting polymer/MWCNT as CE material by dispersing MWCNT in poly(3,4-ethylenedioxy-thiophene):poly(styrenesulfonate) (PEDOT:PSS) and aqueous poly(styrenesulfonateacide) (PSSA) respectively and subjected to ultrasonication to get uniform dispersion. The above-mentioned MWCNT/PEDOT:PSS and MWCNT/PSSA composites were spin-coated on FTO substrates and dried at 110°C for 30 min to form a network-based composite. This method is cost-effective for a reason such as it is based on room temperature and used low-cost solutions. The DSSC with MWCNT/PEDOT:PSS-based CE demonstrated a high PCE of 6.5% much more than what is obtained for MWCNT/PSSA-based CE (3.6%). The high photovoltaic performance by MWCNT/PEDOT:PSS-based CE over MWCNT/PSSA-based CE is because of their composite structure. In MWCNT/PSSA, the MWCNT is covered by the nonconjugated PSSA, which curtails the charge transfer from the redox species to the CE and on the other hand in MWCNT/PEDOT:PSS, the MWCNT is covered by the conductive PEDOT, which will enhance the charge transfer. Therefore MWCNT/PEDOT:PSS-based DSSC performs better [3].

Binder free water-soluble polyelectrolyte-grafted MWCNTs (MWCNT-g-PSSNa) were prepared using a "grafting free" route. The MWCNT-g-PSSNa-based CE was processed by electrostatic spraying the dispersion that gives a highly uniform surface over a large area with an interconnected network consisting of well-exfoliated MWCNTs because it is electrostatic charge-assisted deposition technique. The MWCNT-g-PSSNa CE-based DSSC have a comparable photovoltaic performance to that of Pt-based DSSC. While the device with Pt CE gives a PCE of 8.30%, the device with MWCNT-g-PSSNa CE yields 7.03%. For the MWCNT-g-PSSNa-based device all the other parameters are close to that of Pt including charge transfer resistance, excepting fill factors. This method is very facile as it is operated at low temperature and uses environmentally "friendly" solvent [115].

Luo et al. employed electrophoresis and cyclic voltammetry (CV) techniques in sequence to MWCNT/polythiophene (PTh)-composite-based CE. At first, MWCNTs were acid functionalized and then deposited on FTO by electrophoretic reaction and in the second step PTh was deposited on FTO/MWCNT by CV technique. The total energy conversion efficiency of the DSSC MWCNT/PTh composite film as CE exhibited 4.72% against DSSC with MWCNT CE which showed only 2.68% of efficiency. The enhanced conductivity, catalytic activity, and high surface area of the MWCNT/PTh composite film has increased the PCE by 76.12%. The spherical shape of MWCNT after depositing PTh showed improved roughness (Fig. 7.7a,b), specific surface area and made the electronic transmission capacity of PTh better which all contribute to its high electrocatalytic activity (Fig. 7.8) leading to enhanced photoelectric properties [116].

Yet another successful attempt for making flexible CE was reported by Malara et al. They made a dispersion of MWCNT with dilute polypropylene solution in toluene followed by controlled solvent evaporation yielding the MWCNT/polypropylene composite in powder form which upon compression molding gave thin plates. To remove the residual polymeric capping reactive ion etching treatment with oxygen plasma was employed. They varied the concentration of MWCNT from 5wt% to 20wt% and made four different types. When these flexible MWCNT/polypropylene-composite-based material was used as CE, the solar cell devices yielded a maximum of 6.67% of conversion efficiency for the CE prepared with 20wt% of MWCNT. By morphological and electrochemical studies, it was proved that the excess polymer was selectively removed by oxygen plasma treatment to make CNT into MWCNTs to protrude better out of the polymeric matrix [117].

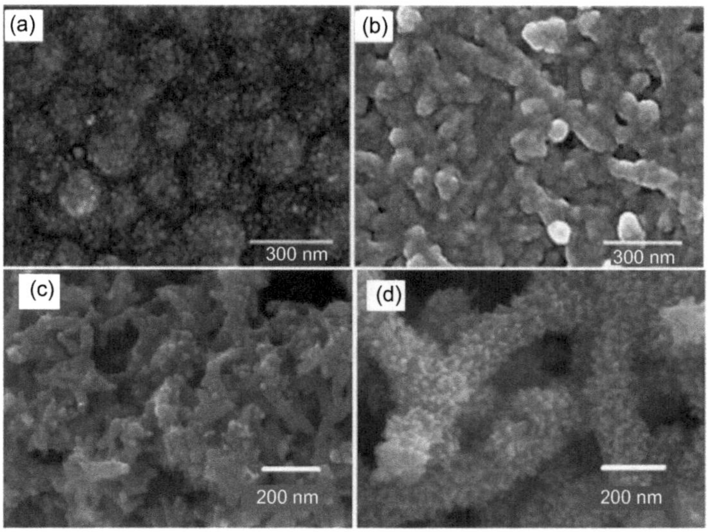

Figure 7.7 SEM images of films of (a) PTh, (b) MWCNT/PTh [116], (c) PANI, and (d) MWCNT/PANI [118].

Niu et al. made a cost-effective MWCNT/polyaniline (PANI) composite by co-polymerizing aniline with carboxylated MWCNTs along with ammonium persulfate in the acidic medium lasting for a different duration. The films comprising MWCNT/PANI was prepared by spin coating the solution on FTO and heat treating it (Fig. 7.7c,d). The photovoltaic performance exhibited by MWCNT/PANI CE is 7.21% which very much close 7.59% by Pt CE [118].

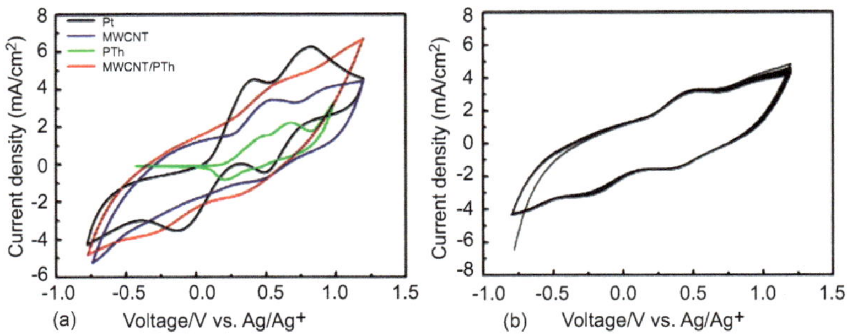

Figure 7.8 CV curves of film electrodes, and (b) consecutive CV curves of MWCNT/PTh electrode [116].

Yue et al. investigated the electrocatalytic properties of MWCNT/polypyrrole (Ppy) composite in the reduction of triiodide to iodide by evaluating the photovoltaic performance of MWCNT/Ppy as CE for DSSC. The composite was prepared by electrochemically from pyrrole electropolymerization system by doping MWCNTs on FTO surface to serve as CE. The surface of the counter electrode had the specific structure of bearing rough surface made of a number of MWCNTs coated on Ppy nanoparticles ensures fast mass transport for the electrolyte and fastening reduction of triiodide to iodide. The composite was prepared for different wt% of MWCNT which ranged from 0.02% to 0.10%. The power conversion measurement for the devices proved that the CE fabricated using the MWCNT/Ppy composite with 0.06 wt% of MWCNTs showed highest PCE of 7.42% and proved to be a better and alternative candidate for Pt which gave only 6.85%. Cyclic voltammetry, electrochemical impedance measurements and Tafel polarization further support the enhanced electrocatalytic activity of the MWCNT/Ppy composite and thus making it a suitable candidate replace Pt. On top of it, this process is cost-effective and can be processed in an ambient atmosphere [119].

7.7 Conclusion

Ever since the discovery of DSSC in 1991, it has attracted researchers around the globe for reasons such as low cost, easy preparation, high efficiency, and being environment-friendly. Till date, researchers have worked on improving three key areas, namely, efficiency, cost, and stability. Although the efforts taken in the past few decades remarkably increased the efficiency of the device to a maximum level to take it forward for commercialization the two factors which hinder commercialization are cost and stability. The counter electrode is an important component of DSSC which determines the photovoltaic conversion and the cost of the device. The most commonly used counter electrode material is Pt as it is highly stable and possesses high catalytic activity to carry out the redox reaction; however, it is an expensive noble metal and this limits the commercialization of DSSC. This drives scientists to explore alternative materials to replace Pt both in terms of catalytic properties and cost. Carbonaceous materials such as carbon nanostructures, activated carbon, CNTs, and graphene have been explored for their suitability to replace Pt. In spite of the drawbacks such as relatively low conductivity, catalytic activity, low adhesion to substrates, and large quantity requirement to meet the required catalytic activity as compared to Pt, carbon-based materials are still preferred because they have appreciable stability, low cost, and need very simple and cost-effective preparation methods. In order to overcome the above-said drawbacks CNTs are made into composites using conductive polymers which are transparent, flexible, easily processed, and have easily tunable properties. Among the polymers, PEDOT has superior properties than Ppy but in terms of cost, it is expensive. PANI-based CEs show better performance with moderate cost factor while PVC and polypropylene-based composites can offer flexible CEs. With further research on incorporating CNTs/polymer-based CEs, the days to make DSSC commercially available at low-cost will not be far away.

Acknowledgments

The corresponding author A. Dennyson Savariraj gratefully acknowledges the FONDECYT postdoctoral Project No. 3170640, Government of Chile, Santiago, for the financial assistance.

References

1. P. Brown, K. Takechi and P. V. Kamat, 2008. Single-walled carbon nanotube scaffolds for dye-sensitized solar cells. *Journal of Physical Chemistry C*, 112(12), 4776–4782.

2. A. Kongkanand, R. M. Dominguez and P. V. Kamat, 2007. Single wall carbon nanotube scaffolds for photoelectrochemical solar cells. Capture and transport of photogenerated electrons. *Nano Letters*, 7(3), 676–680.
3. B. Fan, X. Mei, K. Sun and J. Ouyang, 2008. Conducting polymer/carbon nanotube composite as counter electrode of dye-sensitized solar cells. *Applied Physics Letters*, 93(14), 143103.
4. S. Thomas, T. G. Deepak, G. S. Anjusree, T. A. Arun, S. V. Nair and A. S. Nair, 2014. A review on counter electrode materials in dye-sensitized solar cells. *Journal of Materials Chemistry A*, 2(13), 4474–4490.
5. S. Huang, H. Sun, X. Huang, Q. Zhang, D. Li, Y. Luo and Q. Meng, 2012. Carbon nanotube counter electrode for high-efficient fibrous dye-sensitized solar cells. *Nanoscale Research Letters*, 7(1), 222.
6. J. E. Trancik, S. C. Barton and J. Hone, 2008. Transparent and catalytic carbon nanotube films. *Nano Letters*, 8(4), 982–987.
7. T. N. Murakami, S. Ito, Q. Wang, M. K. Nazeeruddin, T. Bessho, I. Cesar, et al., 2006. Highly efficient dye-sensitized solar cells based on carbon black counter electrodes. *Journal of the Electrochemical Society*, 153(12), A2255–A2261.
8. K. Imoto, K. Takahashi, T. Yamaguchi, T. Komura, J. Nakamura and K. Murata, 2003. High-performance carbon counter electrode for dye-sensitized solar cells. *Solar Energy Materials and Solar Cells*, 79(4), 459–469.
9. L. Kavan, J. H. Yum and M. Gratzel, 2011. Optically transparent cathode for dye-sensitized solar cells based on graphene nanoplatelets. *ACS Nano*, 5(1), 165–172.
10. Z. Huang, X. Liu, K. Li, D. Li, Y. Luo, H. Li, et al., 2007. Application of carbon materials as counter electrodes of dye-sensitized solar cells. *Electrochemistry Communications*, 9(4), 596–598.
11. K. Suzuki, M. Yamaguchi, M. Kumagai and S. Yanagida, 2003. Application of carbon nanotubes to counter electrodes of dye-sensitized solar cells. *Chemistry Letters*, 32(1), 28–29.
12. X. Mei, S. J. Cho, B. Fan and J. Ouyang, 2010. High-performance dye-sensitized solar cells with gel-coated binder-free carbon nanotube films as counter electrode. *Nanotechnology*, 21(39), 395202.
13. W. J. Lee, E. Ramasamy, D. Y. Lee and J. S. Song, 2009. Efficient dye-sensitized solar cells with catalytic multiwall carbon nanotube counter electrodes. *ACS Applied Materials & Interfaces*, 1(6), 1145–1149.

14. S. H. Seo, S. Y. Kim, B. Koo, S. Cha and D. Y. Lee, 2010. Influence of electrolyte composition on the photovoltaic performance and stability of dye-sensitized solar cells with multiwalled carbon nanotube catalysts. *Langmuir*, 26(12), 10341–10346.

15. A. Kay and M. Grätzel, 1996. Low cost photovoltaic modules based on dye sensitized nanocrystalline titanium dioxide and carbon powder. *Solar Energy Materials and Solar Cells*, 44(1), 99–117.

16. P. Ma, N. A. Siddiqui, G. Marom and J. Kim, 2010. Dispersion and functionalization of carbon nanotubes for polymer-based nanocomposites: A review. *Composites Part A: Applied Science and Manufacturing*, 41(10), 1345–1367.

17. S. Park, M. Vosguerichian and Z. Bao, 2013. A review of fabrication and applications of carbon nanotube film-based flexible electronics. *Nanoscale*, 5(5), 1727–1752.

18. D. S. Bethune, C. H. Kiang, M. S. de Vries, G. Gorman, R. Savoy, J. Vazquez, et al., 1993. Cobalt-catalysed growth of carbon nanotubes with single-atomic-layer walls. *Nature*, 363, 605.

19. M. S. Dresselhaus, G. Dresselhaus and R. Saito, 1995. Physics of carbon nanotubes. *Carbon*, 33(7), 883–891.

20. E. T. Thostenson, Z. Ren and T. Chou, 2001. Advances in the science and technology of carbon nanotubes and their composites: A review. *Composites Science and Technology*, 61(13), 1899–1912.

21. B. I. Yakobson and P. Avouris, 2001. Mechanical properties of carbon nanotubes, in *Carbon Nanotubes: Synthesis, Structure, Properties, and Applications* (M. S. Dresselhaus, G. Dresselhaus and P. Avouris, eds), Springer, Berlin, Heidelberg.

22. N. Rouhi, D. Jain and P. J. Burke, 2011. High-performance semiconducting nanotube inks: Progress and prospects. *ACS Nano*, 5(11), 8471–8487.

23. L. Hu, D. S. Hecht and G. Gruner, 2010. Carbon nanotube thin films: Fabrication, properties, and applications. *Chemical Reviews*, 110(10), 5790–5844.

24. Q. Cao and J. A. Rogers, 2009. Ultrathin films of single-walled carbon nanotubes for electronics and sensors: A review of fundamental and applied aspects. *Advanced Materials*, 21(1), 29–53.

25. G. Mittal, V. Dhand, K. Y. Rhee, S. Park and W. R. Lee, 2015. A review on carbon nanotubes and graphene as fillers in reinforced polymer nanocomposites. *Journal of Industrial and Engineering Chemistry*, 21, 11–25.

26. D. Tasis, N. Tagmatarchis, A. Bianco and M. Prato, 2006. Chemistry of carbon nanotubes. *Chemical Reviews*, 106(3), 1105–1136.
27. N. Saifuddin, A. Z. Raziah and A. R. Junizah, 2013. Carbon nanotubes: A review on structure and their interaction with proteins. *Journal of Chemistry*, 676815, 1–18.
28. N. Arora and N. N. Sharma, 2014. Arc discharge synthesis of carbon nanotubes: Comprehensive review. *Diamond and Related Materials*, 50, 135–150.
29. H. Dai, 2002. Carbon nanotubes: Opportunities and challenges. *Surface Science*, 500(1), 218–241.
30. T. W. Ebbesen and P. M. Ajayan, 1992. Large-scale synthesis of carbon nanotubes. *Nature*, 358, 220.
31. G. H. Taylor, J. D. F. Gerald, L. Pang and M. A. Wilson, 1994. Cathode deposits in fullerene formation: Microstructural evidence for independent pathways of pyrolytic carbon and nanobody formation. *Journal of Crystal Growth*, 135(1), 157–164.
32. S. Cui, P. Scharff, C. Siegmund, D. Schneider, K. Risch, S. Klötzer, et al., 2004. Investigation on preparation of multiwalled carbon nanotubes by DC arc discharge under N_2 atmosphere. *Carbon*, 42(5), 931–939.
33. H. Yokomichi, M. Matoba, H. Sakima, M. Ichihara and F. Sakai, 1998. Synthesis of carbon nanotubes by arc discharge in CF_4 gas atmosphere. *Japanese Journal of Applied Physics*, 37(12), 6492.
34. K. Shimotani, K. Anazawa, H. Watanabe and M. Shimizu, 2001. New synthesis of multi-walled carbon nanotubes using an arc discharge technique under organic molecular atmospheres. *Applied Physics A*, 73(4), 451–454.
35. T. Guo, P. Nikolaev, A. Thess, D. T. Colbert and R. E. Smalley, 1995. Catalytic growth of single-walled nanotubes by laser vaporization. *Chemical Physics Letters*, 243(1), 49–54.
36. G. D. Nessim, 2010. Properties, synthesis, and growth mechanisms of carbon nanotubes with special focus on thermal chemical vapor deposition. *Nanoscale*, 2(8), 1306–1323.
37. A. Thess, R. Lee, P. Nikolaev, H. Dai, P. Petit, J. Robert, et al., 1996. Crystalline ropes of metallic carbon nanotubes. *Science*, 273(5274), 483.
38. M. Kumar and Y. Ando, 2010. Chemical vapor deposition of carbon nanotubes: A review on growth mechanism and mass production. *Journal of Nanoscience and Nanotechnology*, 10(6), 3739–3758.

39. Q. Zhao, T. Jiang, C. Li and H. Yin, 2011. Synthesis of multi-wall carbon nanotubes by Ni-substituted (loading) MCM-41 mesoporous molecular sieve catalyzed pyrolysis of ethanol. *Journal of Industrial and Engineering Chemistry*, 17(2), 218–222.

40. A. Govindaraj and C. N. R. Rao, 2002. Organometallic precursor route to carbon nanotubes. *Pure and Applied Chemistry*, 74, 1571.

41. Z. F. Ren, Z. P. Huang, J. W. Xu, J. H. Wang, P. Bush, M. P. Siegal et al., 1998. Synthesis of large arrays of well-aligned carbon nanotubes on glass. *Science*, 282(5391), 1105.

42. K. Lee, W. Yeoh, S. Chai, S. Ichikawa and A. R. Mohamed, 2012. The role of water vapor in carbon nanotube formation via water-assisted chemical vapor deposition of methane. *Journal of Industrial and Engineering Chemistry*, 18(4), 1504–1511.

43. J. S. Kim, Y. Jang and I. Im, 2013. Growth of vertical carbon nanotubes according to the Al_2O_3 buffer layer preparation. *Journal of Industrial and Engineering Chemistry*, 19(5), 1501–1506.

44. S. Huang, M. Woodson, R. Smalley and J. Liu, 2004. Growth mechanism of oriented long single walled carbon nanotubes using "fast-heating" chemical vapor deposition process. *Nano Letters*, 4(6), 1025–1028.

45. S. Patel, Y. Magga, L. Belkady, E. Hibert, D. Porterat, P. Boulanger, et al., 2013. Development and optimization of a secure injection CVD process to grow aligned carbon nanotubes on large substrates. *Journal of Physics: Conference Series*, 429(1), 012053.

46. X. Wang, Q. Li, J. Xie, Z. Jin, J. Wang, Y. Li, et al., 2009. Fabrication of ultralong and electrically uniform single-walled carbon nanotubes on clean substrates. *Nano Letters*, 9(9), 3137–3141.

47. J. Wu, Z. Lan, J. Lin, M. Huang, Y. Huang, L. Fan, et al., 2017. Counter electrodes in dye-sensitized solar cells. *Chemical Society Reviews*, 46(19), 5975–6023.

48. L. Yu, C. Shearer and J. Shapter, 2016. Recent development of carbon nanotube transparent conductive films. *Chemical Reviews*, 116(22), 13413–13453.

49. T. Durkop, S. A. Getty, E. Cobas and M. S. Fuhrer, 2004. Extraordinary mobility in semiconducting carbon nanotubes. *Nano Letters*, 4(1), 35–39.

50. X. Zhou, J. Park, S. Huang, J. Liu and P. L. McEuen, 2005. Band structure, phonon scattering, and the performance limit of single-walled carbon nanotube transistors. *Physical Review Letters*, 95(14), 146805.

51. N. D. Arora, J. R. Hauser and D. J. Roulston, 1982. Electron and hole mobilities in silicon as a function of concentration and temperature. *IEEE Transactions on Electron Devices*, 29(2), 292–295.
52. D. Li, Z. Shao, Q. Hao and H. Zhao, 2014. Intrinsic carrier mobility of a single-layer graphene covalently bonded with single-walled carbon nanotubes. *Journal of Applied Physics*, 115(23), 233701.
53. Y. K. Bahk, X. He, E. Gitsis, Y. Kuo, N. Kim and J. Wang, 2015. Enhanced dispersion stability and mobility of carboxyl-functionalized carbon nanotubes in aqueous solutions through strong hydrogen bonds. *Journal of Nanoparticle Research*, 17(10), 396.
54. L. Hu, D. S. Hecht and G. Grüner, 2009. Infrared transparent carbon nanotube thin films. *Applied Physics Letters*, 94(8), 081103.
55. A. J. Ferguson, A. Dowgiallo, D. J. Bindl, K. S. Mistry, O. G. Reid, N. Kopidakis, et al., 2015. Trap-limited carrier recombination in single-walled carbon nanotube heterojunctions with fullerene acceptor layers. *Physical Review B*, 91(24), 245311.
56. S. V. Morozov, K. S. Novoselov, M. I. Katsnelson, F. Schedin, D. C. Elias, J. A. Jaszczak, et al., 2008. Giant intrinsic carrier mobilities in graphene and its bilayer. *Physical Review Letters*, 100(1), 016602.
57. Y. Ding, V. Singh, S. M. Goodman and P. Nagpal, 2014. Low exciton-phonon coupling, high charge carrier mobilities, and multiexciton properties in two-dimensional lead, silver, cadmium, and copper chalcogenide nanostructures. *Journal of Physical Chemistry Letters*, 5(24), 4291–4297.
58. Z. Yao, C. L. Kane and C. Dekker, 2000. High-field electrical transport in single-wall carbon nanotubes. *Physical Review Letters*, 84(13), 2941–2944.
59. C. Subramaniam, T. Yamada, K. Kobashi, A. Sekiguchi, D. N. Futaba, M. Yumura, et al., 2013. One hundred fold increase in current carrying capacity in a carbon nanotube-copper composite. *Nature Communications*, 4, 2202.
60. J. E. Fischer, H. Dai, A. Thess, R. Lee, N. M. Hanjani, D. L. Dehaas, et al., 1997. Metallic resistivity in crystalline ropes of single-wall carbon nanotubes. *Physical Review B*, 55(8), R4921–R4924.
61. M. in het Panhuis, 2016. Carbon nanotubes: Enhancing the polymer building blocks for intelligent materials, *Journal of Materials Chemistry*, 16, 3598–3605.
62. D. Mann, A. Javey, J. Kong, Q. Wang and H. Dai, 2003. Ballistic transport in metallic nanotubes with reliable Pd ohmic contacts. *Nano Letters*, 3(11), 1541–1544.

63. G. Chen, D. N. Futaba, S. Sakurai, M. Yumura and K. Hata, 2014. Interplay of wall number and diameter on the electrical conductivity of carbon nanotube thin films. *Carbon*, 67, 318–325.

64. Y. Yong, T. Yonezawa, M. Matsubara and H. Tsukamoto, 2015. The mechanism of alkylamine-stabilized copper fine particles towards improving the electrical conductivity of copper films at low sintering temperature. *Journal of Materials Chemistry C*, 3(23), 5890–5895.

65. C. Qiu, Z. Zhang, D. Zhong, J. Si, Y. Yang and L. Peng, 2015. Carbon nanotube feedback-gate field-effect transistor: suppressing current leakage and increasing on/off ratio. *ACS Nano*, 9(1), 969–977.

66. L. Hu, D. S. Hecht and G. Gruner, 2004. Percolation in transparent and conducting carbon nanotube networks. *Nano Letters*, 4(12), 2513–2517.

67. A. Javey, H. Kim, M. Brink, Q. Wang, A. Ural, J. Guo, et al., 2002. High-k dielectrics for advanced carbon-nanotube transistors and logic gates. *Nature Materials*, 1, 241.

68. F. Xia, D. B. Farmer, Y. Lin and P. Avouris, 2010. Graphene field-effect transistors with high on/off current ratio and large transport band gap at room temperature. *Nano Letters*, 10(2), 715–718.

69. E. Pop, D. Mann, Q. Wang, K. Goodson and H. Dai, 2006. Thermal conductance of an individual single-wall carbon nanotube above room temperature. *Nano Letters*, 6(1), 96–100.

70. R. N. Salaway and L. V. Zhigilei, 2014. Molecular dynamics simulations of thermal conductivity of carbon nanotubes: Resolving the effects of computational parameters. *International Journal of Heat and Mass Transfer*, 70, 954–964.

71. P. Kim, L. Shi, A. Majumdar and P. L. McEuen, 2001. Thermal transport measurements of individual multiwalled nanotubes. *Physical Review Letters*, 87(21), 215502.

72. Y. Jiang, E. Juarez-Perez, Q. Ge, S. Wang, M. R. Leyden, L. K. Ono, et al., 2016. Post-annealing of $MAPbI_3$ perovskite films with methylamine for efficient perovskite solar cells. *Materials Horizons*, 3, 548–555.

73. N. A. Katcho, J. Carrete, W. Li and N. Mingo, 2014. Effect of nitrogen and vacancy defects on the thermal conductivity of diamond: An ab initio Green's function approach. *Physical Review B*, 90(9), 094117.

74. M. Yu, O. Lourie, M. J. Dyer, K. Moloni, T. F. Kelly and R. S. Ruoff, 2000. Strength and breaking mechanism of multiwalled carbon nanotubes under tensile load. *Science*, 287(5453), 637.

75. Z. W. Pan, S. S. Xie, L. Lu, B. H. Chang, L. F. Sun, W. Y. Zhou, et al., 1999. Tensile tests of ropes of very long aligned multiwall carbon nanotubes. *Applied Physics Letters*, 74(21), 3152–3154.

76. P. Poncharal, Z. L. Wang, D. Ugarte and W. A. de Heer, 1999. Electrostatic deflections and electromechanical resonances of carbon nanotubes. *Science*, 283(5407), 1513.

77. J. Salvetat, G. A. Briggs, J. Bonard, R. R. Bacsa, A. J. Kulik, T. Stockli, et al., 1999. Elastic and shear moduli of single-walled carbon nanotube ropes. *Physical Review Letters*, 82(5), 944–947.

78. M. M. J. Treacy, T. W. Ebbesen and J. M. Gibson, 1996. Exceptionally high Young's modulus observed for individual carbon nanotubes. *Nature*, 381, 678.

79. D. Khang, J. Xiao, C. Kocabas, S. MacLaren, T. Banks, H. Jiang, et al., 2008. Molecular scale buckling mechanics in individual aligned single-wall carbon nanotubes on elastomeric substrates. *Nano Letters*, 8(1), 124–130.

80. P. Zhou, X. Yang, L. He, Z. Hao, W. Luo, B. Xiong, et al., 2015. The Young's modulus of high-aspect-ratio carbon/carbon nanotube composite microcantilevers by experimental and modeling validation. *Applied Physics Letters*, 106(11), 111908.

81. F. Szuecs, M. Werner, R. S. Sussmann, C. S. J. Pickles and H. J. Fecht, 1999. Temperature dependence of Young's modulus and degradation of chemical vapor deposited diamond. *Journal of Applied Physics*, 86(11), 6010–6017.

82. M. Yu, B. S. Files, S. Arepalli and R. S. Ruoff, 2000. Tensile loading of ropes of single wall carbon nanotubes and their mechanical properties. *Physical Review Letters*, 84(24), 5552–5555.

83. D. A. Walters, L. M. Ericson, M. J. Casavant, J. Liu, D. T. Colbert, K. A. Smith, et al., 1999. Elastic strain of freely suspended single-wall carbon nanotube ropes. *Applied Physics Letters*, 74(25), 3803–3805.

84. N. V. Viet and W. S. Kuo, 2015. Stress distribution in carbon nanotubes with bending fracture. *Engineering Fracture Mechanics*, 136, 158–171.

85. E. W. Wong, P. E. Sheehan and C. M. Lieber, 1997. Nanobeam mechanics: Elasticity, strength, and toughness of nanorods and nanotubes. *Science*, 277(5334), 1971.

86. P. C. Chakraborti, A. Kundu and B. K. Dutta, 2014. Weibull analysis of low temperature fracture stress data of 20MnMoNi55 and SA333 (Grade 6) steels. *Materials Science and Engineering A*, 594, 89–97.

87. M. Cinke, J. Li, B. Chen, A. Cassell, L. Delzeit, J. Han, et al., 2002. Pore structure of raw and purified HiPco single-walled carbon nanotubes. *Chemical Physics Letters*, 365(1), 69–74.

88. K. H. Kim, Y. Oh and M. F. Islam, 2013. Mechanical and thermal management characteristics of ultrahigh surface area single-walled carbon nanotube aerogels. *Advanced Functional Materials*, 23(3), 377–383.

89. O. Barbieri, M. Hahn, A. Herzog and R. Kötz, 2005. Capacitance limits of high surface area activated carbons for double layer capacitors. *Carbon*, 43(6), 1303–1310.

90. D. Li, Y. Tian, Y. Qiao and L. Wen, 2014. Conversion of powdered active carbon into monoliths without reducing specific surface area using H_3PO_4-impregnated waste sawdust. *Materials Letters*, 125, 175–178.

91. F. Du, J. E. Fischer and K. I. Winey, 2003. Coagulation method for preparing single-walled carbon nanotube/poly(methyl methacrylate) composites and their modulus, electrical conductivity, and thermal stability. *Journal of Polymer Science Part B: Polymer Physics*, 41(24), 3333–3338.

92. M. S. P. Shaffer and A. H. Windle, 1999. Fabrication and characterization of carbon nanotube/poly(vinyl alcohol) composites. *Advanced Materials*, 11(11), 937–941.

93. B. Safadi, R. Andrews and E. A. Grulke, 2002. Multiwalled carbon nanotube polymer composites: synthesis and characterization of thin films. *Journal of Applied Polymer Science*, 84(14), 2660–2669.

94. J. Jang, J. Bae and S. Yoon, 2003. A study on the effect of surface treatment of carbon nanotubes for liquid crystalline epoxide-carbon nanotube composites. *Journal of Materials Chemistry*, 13(4), 676–681.

95. X. Gong, J. Liu, S. Baskaran, R. D. Voise and J. S. Young, 2000. Surfactant-assisted processing of carbon nanotube/polymer composites. *Chemistry of Materials*, 12(4), 1049–1052.

96. S. Cui, R. Canet, A. Derre, M. Couzi and P. Delhaes, 2003. Characterization of multiwall carbon nanotubes and influence of surfactant in the nanocomposite processing. *Carbon*, 41(4), 797–809.

97. R. Sen, B. Zhao, D. Perea, M. E. Itkis, H. Hu, J. Love, et al., 2004. Preparation of single-walled carbon nanotube reinforced polystyrene and polyurethane nanofibers and membranes by electrospinning. *Nano Letters*, 4(3), 459–464.

98. Y. Dror, W. Salalha, R. L. Khalfin, Y. Cohen, A. L. Yarin and E. Zussman, 2003. Carbon nanotubes embedded in oriented polymer nanofibers by electrospinning. *Langmuir*, 19(17), 7012–7020.

99. R. Andrews, D. Jacques, D. Qian and T. Rantell, 2002. Multiwall carbon nanotubes: Synthesis and application. *Accounts of Chemical Research*, 35(12), 1008–1017.

100. D. E. Hill, Y. Lin, A. M. Rao, L. F. Allard and Y. Sun, 2002. Functionalization of carbon nanotubes with polystyrene. *Macromolecules*, 35(25), 9466–9471.

101. J. Y. Kim and S. H. Kim, 2006. Influence of multiwall carbon nanotube on physical properties of poly(ethylene 2,6-naphthalate) nanocomposites. *Journal of Polymer Science Part B: Polymer Physics*, 44(7), 1062–1071.

102. R. Haggenmueller, W. Zhou, J. E. Fischer and K. I. Winey, 2003. Production and characterization of polymer nanocomposites with highly aligned single-walled carbon nanotubes. *Journal of Nanoscience and Nanotechnology*, 3(1–2), 105–110.

103. M. Moniruzzaman and K. I. Winey, 2006. Polymer nanocomposites containing carbon nanotubes. *Macromolecules*, 39(16), 5194–5205.

104. D. Qian, E. C. Dickey, R. Andrews and T. Rantell, 2000. Load transfer and deformation mechanisms in carbon nanotube-polystyrene composites. *Applied Physics Letters*, 76(20), 2868–2870.

105. H. J. Yoo, Y. C. Jung, N. G. Sahoo and J. W. Cho, 2006. Polyurethane-carbon nanotube nanocomposites prepared by in-situ polymerization with electroactive shape memory. *Journal of Macromolecular Science, Part B*, 45(4), 441–451.

106. T. Wu and S. Lin, 2006. Synthesis, characterization, and electrical properties of polypyrrole/multiwalled carbon nanotube composites. *Journal of Polymer Science Part A: Polymer Chemistry*, 44(21), 6449–6457.

107. M. Kang, S. J. Myung and H. Jin, 2006. Nylon 610 and carbon nanotube composite by in situ interfacial polymerization. *Polymer*, 47(11), 3961–3966.

108. M. Moniruzzaman, J. Chattopadhyay, W. E. Billups and K. I. Winey, 2007. Tuning the mechanical properties of SWNT/Nylon 6,10 composites with flexible spacers at the interface. *Nano Letters*, 7(5), 1178–1185.

109. H. J. Choi, J. E. Shin, G. Lee, N. Park, K. Kim and S. C. Hong, 2010. Effect of surface modification of multi-walled carbon nanotubes on the fabrication and performance of carbon nanotube based counter electrodes for dye-sensitized solar cells. *Current Applied Physics*, 10(2), S165–S167.

110. J. Park, M. S. Akhtar, Z. Y. Li, D. Cho, W. Lee and O. Yang, 2012. Application of single walled carbon nanotubes as counter electrode for dye sensitized solar cells. *Electrochimica Acta*, 85, 600–604.

111. J. Velten, A. J. Mozer, D. Li, D. Officer, G. Wallace, R. Baughman, et al., 2012. Carbon nanotube/graphene nanocomposite as efficient counter electrodes in dye-sensitized solar cells. *Nanotechnology*, 23(8), 085201.

112. J. G. Nam, Y. J. Park, B. S. Kim and J. S. Lee, 2010. Enhancement of the efficiency of dye-sensitized solar cell by utilizing carbon nanotube counter electrode. *Scripta Materialia*, 62(3), 148–150.

113. J. Zhang, X. Li, W. Guo, T. Hreid, J. Hou, H. Su, et al., 2011. Electropolymerization of a poly(3,4-ethylenedioxythiophene) and functionalized, multi-walled, carbon nanotubes counter electrode for dye-sensitized solar cells and characterization of its performance. *Electrochimica Acta*, 56(9), 3147–3152.

114. S. G. Hashmi, J. Halme, Y. Ma, T. Saukkonen and P. Lund, 2014. A single-walled carbon nanotube coated flexible PVC counter electrode for dye-sensitized solar cells. *Advanced Materials Interfaces*, 1(2), 1300055.

115. J. Han, H. Kim, D. Y. Kim, S. M. Jo and S. Jang, 2010. Water-soluble polyelectrolyte-grafted multiwalled carbon nanotube thin films for efficient counter electrode of dye-sensitized solar cells. *ACS Nano*, 4(6), 3503–3509.

116. J. Luo, H. Niu, W. Wu, C. Wang, X. Bai and W. Wang, 2012. Enhancement of the efficiency of dye-sensitized solar cell with multi-wall carbon nanotubes/polythiophene composite counter electrodes prepared by electrodeposition. *Solid State Sciences*, 14(1), 145–149.

117. F. Malara, M. Manca, L. De Marco, P. Pareo and G. Gigli, 2011. Flexible carbon nanotube-based composite plates as efficient monolithic counter electrodes for dye solar cells. *ACS Applied Materials & Interfaces*, 3(9), 3625–3632.

118. H. Niu, S. Qin, X. Mao, S. Zhang, R. Wang, L. Wan, et al., 2014. Axle-sleeve structured MWCNTs/polyaniline composite film as cost-effective counter-electrodes for high efficient dye-sensitized solar cells. *Electrochimica Acta*, 121, 285–293.

119. G. Yue, L. Wang, X. Zhang, J. Wu, Q. Jiang, W. Zhang, et al., 2014. Fabrication of high performance multi-walled carbon nanotubes/polypyrrole counter electrode for dye-sensitized solar cells. *Energy*, 67, 460–467.

120. G. Aragay, J. Pons, A. Merkoci, 2011. Enhanced electrochemical detection of heavy metals at heated graphite nanoparticle-based screen-printed electrodes, *Journal of Materials Chemistry*, 21, 4326–4331.

Chapter 8

Fabrication of Carbon Nanofibers Based Composites for High Performance Dye-Sensitized Solar Cells

E. Vivek
Department of Physics, University College of Engineering, Anna University (BIT Campus), Tiruchirappalli, Tamil Nadu, India 620024
vivek.nanoscience@gmail.com

Carbon nanofiber (CNF) is one of the most significant classes of the carbon family, which has been investigated in a wide range of potential applications. They hold sp^2 hybridization-based linear discontinuous filaments and are highly graphitic. This chapter discusses and summarizes the structure, preparation methods, and catalytic use of the material and its composite as a counter electrode (CE) in the dye-sensitized solar cell (DSSC). Carbon nanofibers contain well-organized graphitic layers in which the orientation is respect to the axis of the fiber this can be parallel, perpendicular, or stacked, which is known to be the herringbone structure. The CNF can be prepared by two major techniques, which are chemical vapor deposition (CVD) and electrospinning. In this perspective, the catalytic behavior of CNF

Counter Electrode for Dye-Sensitized Solar Cells
Edited by Alagarsamy Pandikumar, Ramesh Mohan, and Kandasamy Jothivnekatachalam

ISBN 978-981-4877-38-1 (Hardcover), 978-1-003-11077-4 (eBook)
www.jennystanford.com

and its composites as a counter electrode in DSSC are discussed with the detailed review based on the reported observations. Finally, this chapter concludes with the prospects and challenges of CNF-based counter electrodes and their application in DSSCs in future.

8.1 Introduction

Dye-sensitized solar cells are considered as a potential candidate for an alternative to the conventional silicon-based solar cells and organic photovoltaic devices due to their low cost, ease of fabrication, and high power conversion efficiency [1]. DSSCs consists of n-type nanoporous metal oxide semiconductors such as TiO_2 and ZnO nanoparticles (NPs), where the surface would be sensitized by a sensitizer (e.g., N917 dye, which absorbs the sunlight), iodine (I^-/I_3^-) electrolyte, and platinum (Pt) counter electrode as a catalyst. The counter electrode is one of the major component that collects the electron from the external circuit and reduces the electrolyte from triiodide (I_3^-) to iodide (I^-), which regenerates the sensitizer and acts as a catalyst. The performance of the DSSC is greatly dependent on the suppression of electron–hole pair recombination at the photoanode, which is mainly influenced by the fast reduction reaction at the counter electrode. Consequently, by considering the importance of the ideal role of the counter electrode for efficient DSSC, the counter electrode material is expected to have an outstanding catalytic activity, good electrical conductivity, and chemical stability. Pt is generally used as the counter electrode material, but the high cost and the corrosiveness to the electrolyte limit the long-term stability of the device and obstruct its practical application. So the investigations toward the carbonaceous material are still in progress. This chapter discusses and summarizes the structure, properties, preparation methods, and catalytic use of CNF and its composite as a counter electrode material in the DSSC.

The history of carbon nanofibers dates back to 1889 when a patent published by Hughes and coworker reported the unintentional growth of carbon filaments from the carbon-containing gases using a metallic crucible [2]. Robertson was one who first recognized that the interaction between the methane and the metal surface leads to the growth of graphitic carbon at low temperature [3]. For more than 80 years of the century, the carbon nanofibers have been often referred to as carbon filaments and also considered as nuisance due to the occurrence of filamentous carbon in metallic catalysts used in the process of Fischer–Tropsch or steam–methane reforming reactions. In the last two decades, carbon nanofibers have proven to be a freestanding robust individual structure with high electron transport and corrosion resistivity, which are significant for the functionalization of materials and their technology use in aerospace, construction, sports, electronic device, electrochemical devices, and automobile industries [4–7]. After 40 years of continuous efforts, the potential of using carbon nanofiber-based products has grown up to about 12,500 tons/year in the international market.

8.2 Structure and Properties of CNF

CNFs are well-defined one-dimensional fibrous nanostructures. Their diameters range from few nanometers up to several centimeters, and they also hold sp^2 hybridization-based linear discontinuous filaments and are highly graphitic. CNFs consist of ordered graphene layers in which the structure of CNF is presented by the dependence of angle between the fiber axis and the graphene layer. The internal structure of CNFs has been observed by the cross-sectional TEM imaging [8].

- Platelet/stacked
- Herringbone/cup stacked

Platelet structures are described as graphene layers stacked perpendicular to the fiber axis as shown in Fig. 8.1c. The second most usually recognized structural configurations are "herringbone-type," shown in Fig. 8.1f, in which the conical graphene layers, when viewed in the cross section, resemble a "fishbone" and graphene layers are stacked at an angle α of ~10–45° with the nanofiber axis. Additionally, hollow CNFs can also be produced when the angle becomes very steep and it is nearly parallel to the CNFs axis (Fig. 8.1e). The electrical and mechanical property of the vapor-grown carbon fiber at 1100°C measured before graphitization was 2×10^{-3} Ω/cm and after graphitization, it decreased up to 5×10^{-5} Ω/cm. The tensile strength and modulus of the carbon nanofiber were 2.9 GPa and 240 GPa, respectively. It is important to note that the surface chemistry and physical properties of the nanofibers are mainly dependent on the number of layers, cone angle, defect density, and so on.

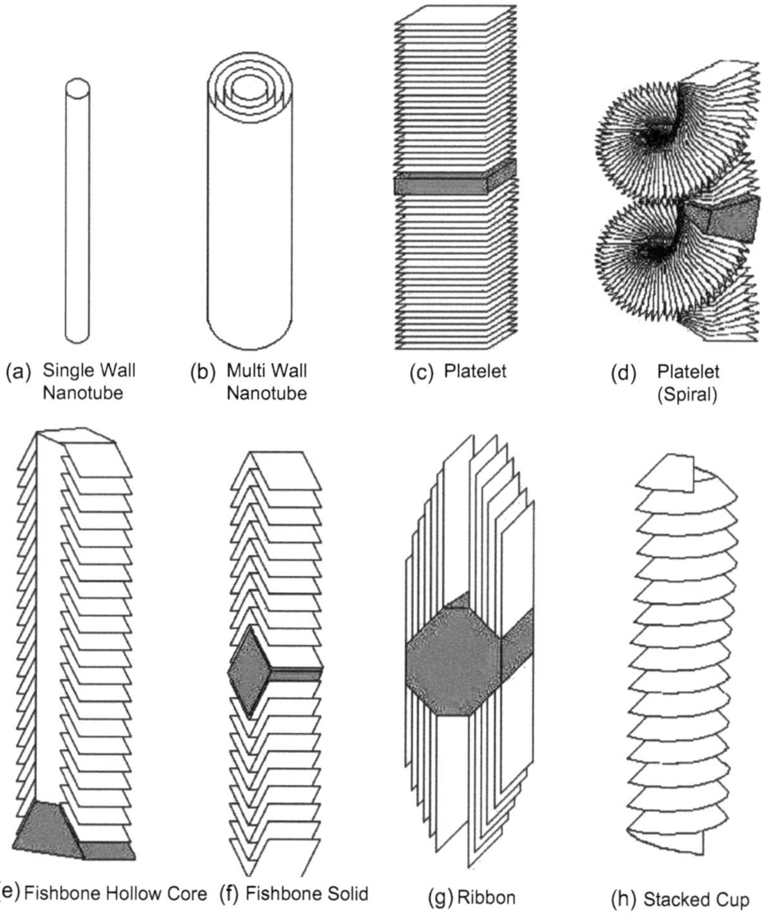

Figure 8.1 Schematic diagram of the different accepted structures of CNT (carbon nanotube) and CNF. Reprinted from Ref. [12], Copyright 2006, with permission from Elsevier.

8.3 Synthesis

Carbon nanofibers can be synthesized by some common methods, such as arc-discharge method, chemical vapor deposition technique, hot filament-assisted sputtering, and electrospinning technique.

8.3.1 Arc-Discharge Method

A method for the synthesis of carbon nanofibers is the arc-discharge method. In general, the working principle of the method is as follows. The setup for the growth of carbon nanofibers using the arc-discharge method is shown in Fig. 8.2. A DC arc voltage is applied between the two graphite electrodes (anode and cathode) in an inert gas atmosphere such as helium. Arc discharge is achieved through the electrical breakdown of the gas, which produces the plasma discharge between them. The factors influencing the product formation are pressure of the inert gas, voltage, current, and the electrode material. Fullerenes are deposited inside the chamber as soot and multi-walled carbon nanotubes at the cathode, while pure graphite rods are used as the electrode material. The single-walled carbon nanotubes are also generated in the form of soot, while using the metal catalyst (Fe or Co) contained graphite as anode and pure graphite as cathode. In a typical synthesis, stacked graphene sheets at various angled CNFs are produced by the arc-discharge method. The carbon rod is the anode with the size of 15 × 15 mm^2 where the cross section is doped with Fe, Co, and NiS (1:1:8 in atomic ratio, 3.0 wt% in total), and the graphite bowl is the cathode where the

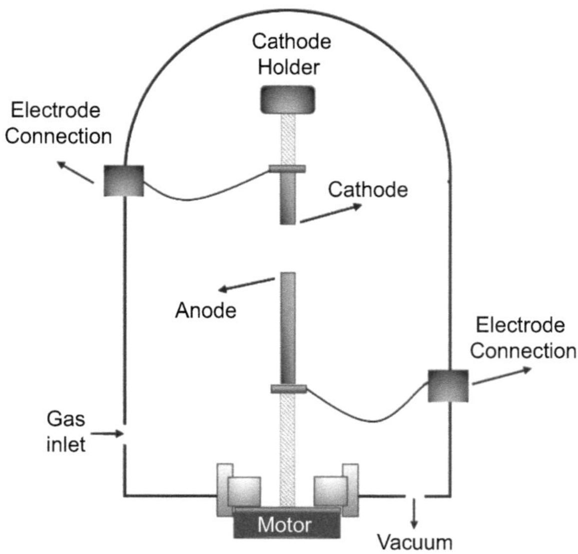

Figure 8.2 Schematic drawing of the setup for arc-discharge method.

small holes are filled with the mixture of FeS, CoS, and NiS with an atomic ratio of 1:1:8. Then the arc discharge is carried out in the helium (>99.999%) atmosphere at a pressure of 8×10^4 Pa. The DC arc current is maintained at 180 A for 30 min. The deposited soot of carbon nanofibers is collected from the cathode surface after the arc discharge.

8.3.2 Chemical Vapor Deposition Technique

The most established method for the production of CNFs is the CVD method. The schematic diagram of CVD instrument is shown in Fig. 8.3a. In a typical CVD process, several volatile precursors react and decompose over the exposed substrate to produce thin film, powder, or single crystal. Materials with a wide range of physical and chemical properties can be deposited by varying the experimental conditions, such as substrate temperature, composition of the reaction gas mixture, and total pressure gas flows. The significant features of the CVD technique are its uniform thickness coatings, outstanding throwing power, and capability of selective deposition. Thin film applications of CVD include passivation layers, conductive oxides, tribological, conductors, corrosion and heat-resistant coatings, dielectrics, and growth of epitaxial layers.

In the early 1980s, Tibbetts' group [10] and Endo and coworkers [11] developed a synthesis method for carbon nanofibers using the CVD technique. The vapor-grown process is the most common process for producing CNF by the CVD method. Figure 8.3b shows the step-by-step nucleation and growth of VGCNFs. The first step is decomposition of methane into carbon and hydrogen atoms over the surface of nickel. The second step is the formation of nickel carbide by occurrence of desorption of H_2 molecules. The third step involves the decomposition of nickel carbide into nickel metal and graphite. The fourth step involves the formation of graphitic layers when the nickel metal squeezed out due to pressure. The fifth step is the growth of VGCNFs which can be occurred continuously through the repetition process of the above four steps. Iron (Fe), nickel (Ni), cobalt (Co), and copper (Cu) are transition metal catalysts used as aggregation with a carbon source gas such as carbon monoxide or a hydrocarbon (methane, ethylene, acetylene, carbon monoxide, and aromatics) at a temperature range of 500–1200°C and high partial pressures of hydrogen [12]. The characteristics and morphology of CNF depend on the catalyst, feedstock, and operating conditions. Platelet nanofibers at low temperature and tubular fibers at higher temperature are yields from Fe catalyst. On the other hand, herringbone nanofibers are obtained from the Ni catalyst or a bimetallic Ni–Fe. Platelet nanofibers may also result from Ni catalyst by maintaining specific conditions, such as large particle size and temperature above 900 K.

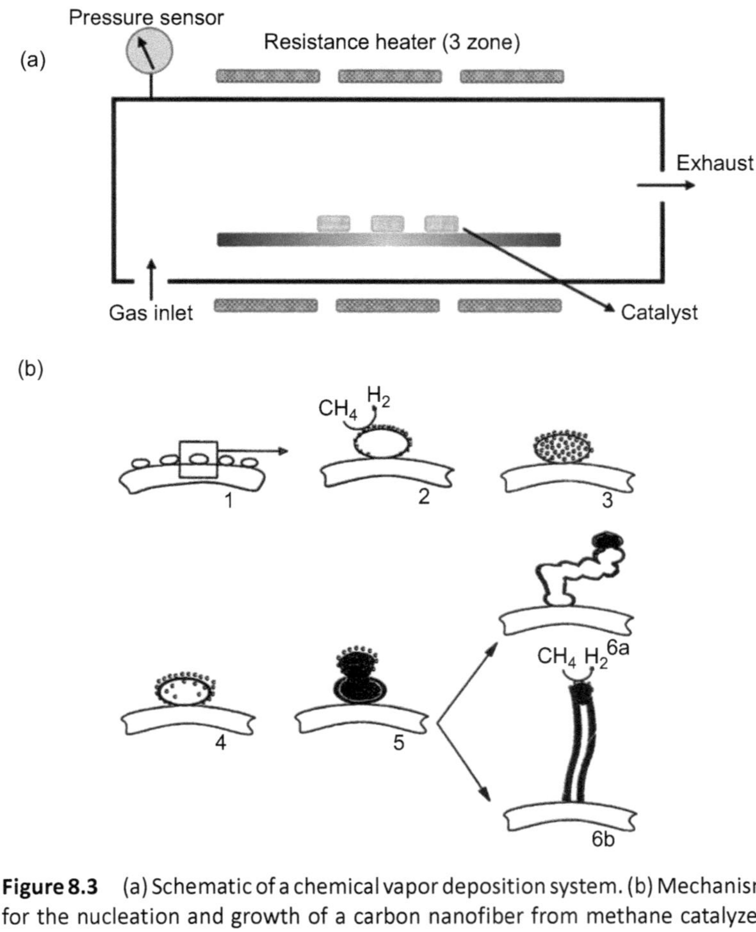

Figure 8.3 (a) Schematic of a chemical vapor deposition system. (b) Mechanism for the nucleation and growth of a carbon nanofiber from methane catalyzed by a supported metal particle. Reprinted from Ref. [9], Copyright 2000, with permission from Elsevier.

8.3.3 Electrospinning

Another route to the fabrication of CNFs is through electrospinning [13]. The schematic representation of electrospinning process is shown in Fig. 8.4. The setup comprises a high voltage power supply (usually in the kV range), an electrically conductive polymer solution, an electrode collector, and a syringe needle and a grounded collector. In a typical process, a high voltage is applied to the tip of the needle where the droplet of the solution starts to overcome the surface tension of the fluid due to the applied electrostatic charge and forms a conical droplet called Taylor cone. Then the electrical field draws the droplet to the opposite electrode (collector) and the fibrous structure is collected. For the preparation of CNFs, polymeric precursors such as polyacrylonitrile (PAN), cellulose, and pitch are used. PAN is the most desirable precursor and regularly used due to its ease of fabrication, mass production, high carbon yield, and strength [14, 15]. The electrospun PAN nanofibers will undergo pyrolysis in an inert gas atmosphere to form CNFs [16–18].

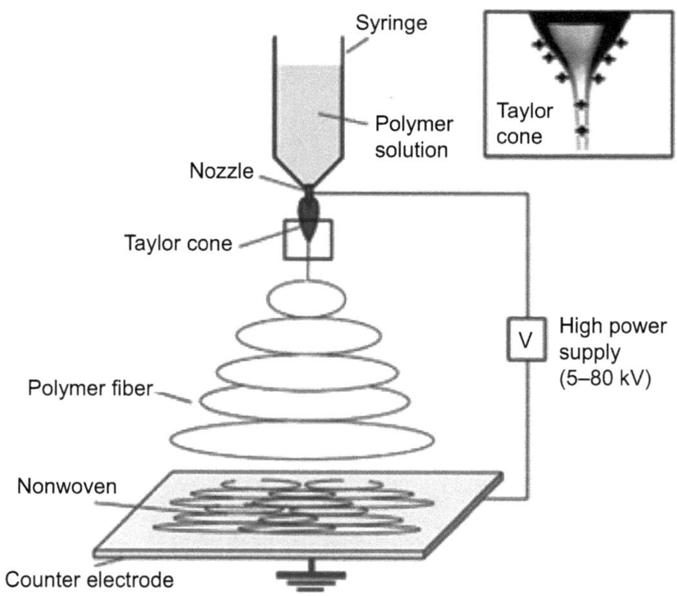

Figure 8.4 Schematic diagram of electrospinning technique. Reprinted from Ref. [55], Copyright 2019, with permission from Elsevier.

8.4 Carbon Nanofibers for Counter Electrode

Stacking morphology and defect-rich edges make CNFs a more promising candidate for the electrochemical activity [19–22]. The CV curve of CNF in Fig. 8.5a shows two pairs of redox peaks and exhibits an electrocatalytic feature that is similar to that of Pt counter electrode [23]. For the first time, Joshi et al. studied the electrospun carbon nanofiber (ECN) as a low-cost alternative to Pt counter electrode [23]. In terms of short-circuit current density (J_{sc}) and open-circuit voltage (V_{oc}), it shows similar performance. Under the AM 1.5 illumination at 100 mW/cm^2, ECN-based cells exhibit energy conversion efficiency (η) of 5.5%. The low cell performance is due to the low fill factor (FF) of the ECN-based cells, which might be caused by higher total series resistance (R_{Stot}) (~15.5 Ω cm^2) of the ECN-based cells than that of the Pt-based devices (~4.8 Ω cm^2). Veerappan et al. reported the use of CNFs with antler and herringbone structures as CEs in DSSCs [24]. Figure 8.5b shows the schematic diagram of the catalytic sites for (I_3^-) reduction in various structures, such as hollow CNT, herringbone carbon nanofiber (CNF-100), and antler carbon nanofiber (CNF-LSA). And it has a faster triiodide reduction rate and a lower charge transfer resistance (R_{ct}) of ~0.5 Ω cm^2 than Pt (~2.3 Ω cm^2). Also, the glass-based devices showed energy conversion efficiencies of 7.0% and 5.0% for the flexible cells with the quasi-solid state electrolyte.

Through concentric electrospinning, Park et al. prepared activated carbon nanofibers with hollow core/highly mesoporous shell structure (meso-HACNF) as a novel counter electrode (Fig. 8.5c) [25]. Meso-HACNF with 10 wt% PMMA shows a high efficiency of 7.21%, which is due to the large surface area of about 1191 m^2/g with mesoporous surface area of 151 m^2/g. The large surface area of the electrode material increases the contact between the electrode and the electrolyte which results in the reduction of charge transfer resistance. Park et al. also prepared a hollow activated carbon nanofibers (HACNF) using the same concentric electrospinning method (Fig. 8.5d) [26]. The surface area of HACNF is 1037.5 m^2/g and the overall parameters of HACNF show J_{sc} of 14.5 mA/cm^2, V_{oc} of 0.73 V, FF of 62%, efficiency (η) of 6.58%. Chen and

Hsieh reported the synthesis of direct growth of graphitic nanofibers on fluorine-doped tin oxide (FTO) glass as a counter electrode for dye-sensitized solar cells by using oxygen-assisted low-pressure chemical vapor deposition (LPCVD) method at a low temperature (550°C) [27].

Table 8.1 Summary of photovoltaic performance of DSSCs fabricated using CNF-based counter electrode

Counter electrode	J_{sc} (mA/cm^2)	V_{oc} (V)	FF	Efficiency (η, %)	Reference
ECN	12.60	0.76	57	5.5	[23]
CNF-LSA	12.1	0.830	69.7	7.0	[24]
CNF-100	12.0	0.826	69.7	6.8	
Meso-HACNF1	15.4	0.73	64	7.21	[25]
Meso-HACNF2	15.3	0.74	61	6.91	
HACNF	14.5	0.73	62	6.58	[26]
Dense platelet GNFs	10.50	0.81	65	5.51	[28]
HCNF	15.51	0.701	62	6.74	[29]

In combination with the photoanode, electrolyte, and counter electrode (graphitic nanofibers) under AM 1.5 (100 mW/cm^2) illumination, the cells showed a power conversion efficiency of 5.51%. Wang et al. investigated the helical carbon nanofiber (HCNF) as the low-cost counter electrode for dye-sensitized solar cells [28]. The electrochemical studies indicate that R_{ct} of HCNF is 1.1 Ω cm^2 in the (I^-/ I_3^-) redox electrolyte. These fabricated dye-sensitized solar cells achieve an energy conversion efficiency of 6.74%. Flexible CNF mats have been successfully fabricated by Zhao et al. via electrospinning [29]. The specific surface area is about 941 m^2/g and its maximum power conversion efficiency is up to 7.60%. The photovoltaic performance of CNF-based counter electrode is summarized in Table 8.1. These reports suggest that the cost-effectiveness and flexibility make CNF a suitable and promising candidate to replace the conventional and expensive platinum in DSSCs.

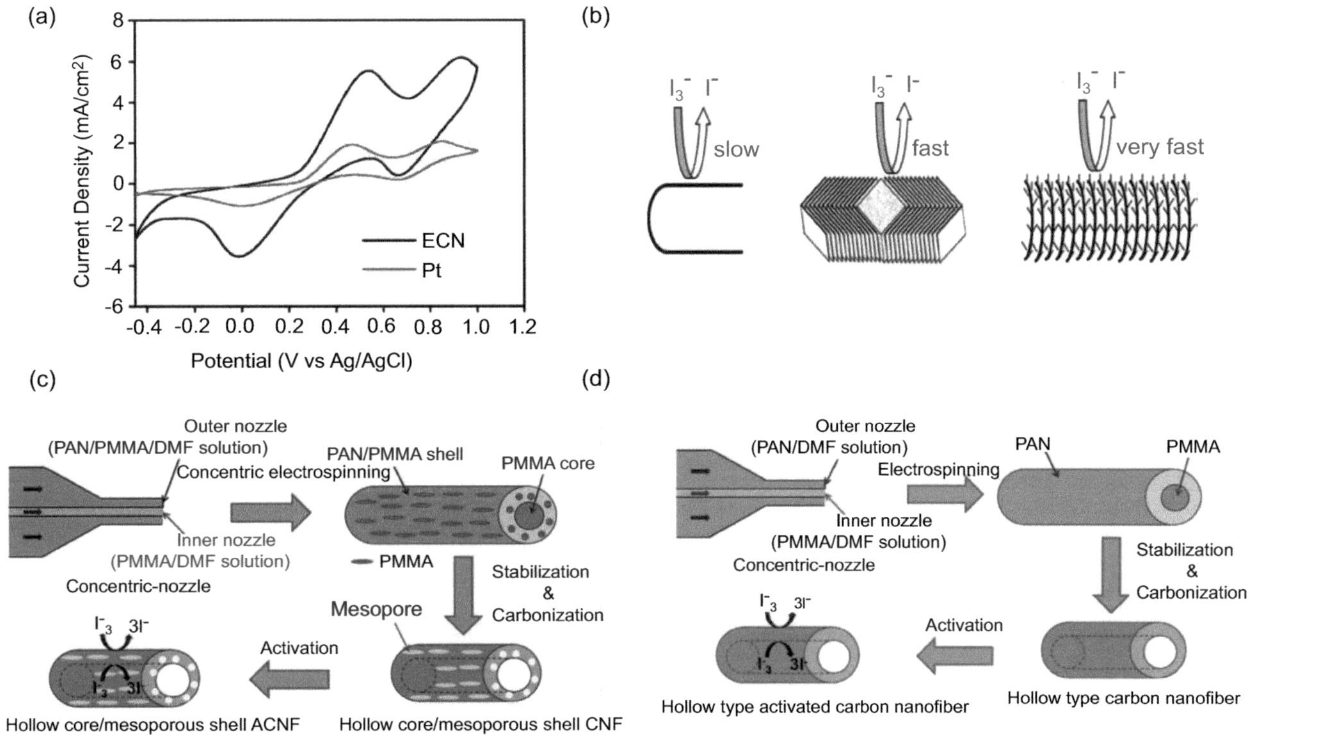

Figure 8.5 (a) Cyclic voltammograms of the ECN (black) and Pt (red) counter electrode. (b) Schematic diagram representing (I_3^-) reduction and oxidation on different fiber surfaces. (c) The fabricating process and (I_3^-) reduction of meso-HACNF. (d) The fabricating process and (I_3^-) reduction of hollow type activated carbon nanofiber. Figure (a) reprinted from Ref. [23], Copyright 2010, with permission from ACS Publications, and Figs. (b–d) reprinted from Refs. [24], Copyright 2011; Ref. [25], Copyright 2013; and Ref. [26], Copyright 2013; respectively, with permission from Elsevier.

8.5 Composite Electrode

To further improve the performance of DSSCs, researchers adopted a new strategy called CNF-based composite counter electrodes. These composite materials are made of two or more different materials often to generate unique properties for replacing the conventional Pt electrode. Poudel et al. demonstrated a composite electrode for the first time by using ECNs and Pt nanoparticles [30]. ECN–Pt composite exhibits an efficient electrocatalyst with lower R_{ct}, larger surface area, and faster reaction rate for the reduction of I_3^- than that of the Pt electrode. The energy conversion efficiency of ECN–Pt-based composite is 8%.

Joshi et al. reported a nickel incorporated carbon nanotube/nanofiber composite (Ni-CNT-CNF) as a low-cost alternative to Pt CE for DSSCs [31]. The R_{ct} of Ni-CNT-CNF measured by using electrochemical impedance spectroscopy (EIS) was 0.71 Ω cm^2, which is lower than that of the Pt electrode (1.81 Ω cm^2). The Ni-CNT-CNF composite CEs produced an efficiency of 7.96% with a J_{sc} of 15.83mA/cm^2, V_{oc} of 0.80 V, and FF of 0.63. Barakat et al. tried a bimetallic composite of Pd–Co-doped carbon nanofibers as the counter electrode [32]. The energy conversion efficiency is 2.5%, which is lower than the conventional Pt, but the author aims to introduce the study in the perspective of low-cost bimetallic counter electrode. Mahpeykar et al. studied the vertically aligned carbon nanofibers (VACNFs) grown on FTO substrate by low-temperature AC–DC PECVD as the counter electrode for DSSC (Fig. 8.6) [33]. They compared the two types of counter electrode, which are VACNFs grown on FTO-coated glass and hybrid VACNF/Pt arrays on FTO-coated glass, with Pt nanoparticle counter electrodes. The VACNF shows an energy conversion efficiency of 5.6% as compared to 6.5% of Pt counter electrode. VACNF/Pt hybrid counter electrodes delivered an energy conversion efficiency of 7.2%. This increased efficiency is due to the combination of excellent charge transport and catalytic ability of VACNFs and high conductivity of Pt nanoparticles in hybrid electrodes.

Yousef et al. reported CuNi bimetallic nanoparticles incorporated in carbon nanofiber prepared by the low-cost facile electrospinning technique [34]. The prepared nanofibers show a power conversion efficiency of 3.5%, open-circuit voltage of 0.70 V, and a short-circuit current density 7.67 mA/cm^2. Thapa et al. demonstrated series of counter electrodes in different compositions of carbon nanofibers (ECN) and carbon nanoparticles (CNPs) [35]. In that, the ECN:CNP (30:70) showed J_{sc} of 13.37 mA/cm^2, V_{oc} of 0.79 V, FF of 62.8%, and efficiency of 6.64%. Sigdel et al. studied the spray-coated electrospun carbon nanofibers/TiO_2 (1:1) on a FTO substrate as CEs for DSSCs [36]. The ECN/TiO_2 CEs showed η of 7.25%, which is comparable to that of thermally deposited Pt-based DSSCs (η = 7.57%). Jeong et al. demonstrated the spray-coated film of WC/C-NF on FTO glass substrate, which formed the 3D porous network [37]. The schematic illustration is shown in Fig. 8.7. The author compared the cell behavior with two electrolytes I^-/I_3^- and T_2/T^-. As a result, PCE for WC/C-NF CE exhibited 7.77% for I_3^- electrolyte and 5.85% for T_2/T^-.

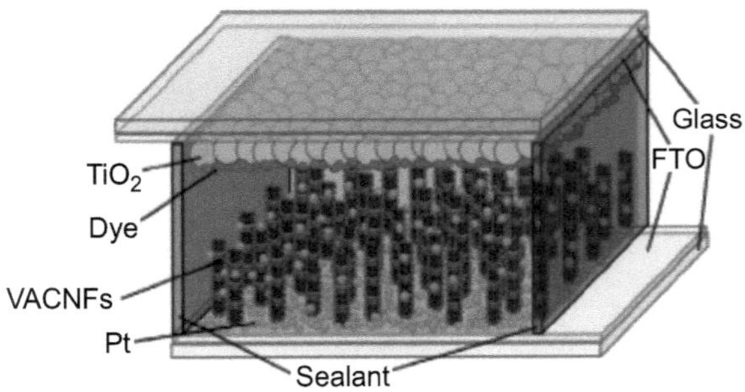

Figure 8.6 Schematic diagram of device assemble with hybrid VACNF/Pt arrays as the counter electrode material. Reprinted from Ref. [33], Copyright 2013, with permission from IOP Publishing Ltd.

An et al. studied nitrogen-doped TiO_2 nanoparticle–CNF composites as counter electrode in DSSCs [38]. The prepared total nitrogen content in TiO_2 nanoparticles was 13.8 wt%. The fabricated device showed the current density of 15.65 mA/cm^2, FF of 60.50%, and power conversion efficiency of 6.31%. Aboagye et al. developed a controlled growth of Pt nanoparticles over the surface of electrospun carbon nanofiber through redox reaction as a counter electrode material for DSSCs [39]. The group investigated three different loading amounts of Pt in ECNF in which ECNFs/PtNPs-3 shows the R_{ct} value of 0.52 Ω cm^2, which is highly catalytic and

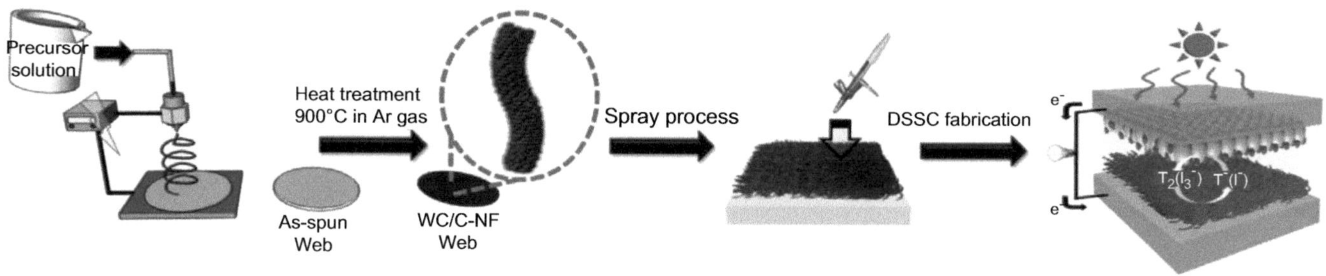

Figure 8.7 Schematic illustrations for the synthesis of WC/C-NF and the device fabrication process as a counter electrode of DSSC. Reprinted from Ref. [37], Copyright 2014, with permission from Elsevier.

exhibits the cell characteristics, such as V_{oc} of 0.83 V, J_{sc} of 13.33 mA/cm^2, FF of 0.688, and η of 7.6%. Motlak et al. reported the synthesis of NiCo nanoparticle-doped carbon nanofibers and their use as counter electrode materials for DSSCs [40]. The results showed that the photovoltaic conversion efficiency of NiCo nanoparticles doped carbon nanofibers counter electrode is 4.47%. Saranya et al. studied the titanium carbide (TiC) embedded CNFs by the electrospinning technique [41]. The DSSC fabricated with counter electrode TiC (10 wt%) showed a photovoltaic performance of V_{oc} of 0.72 V, J_{sc} of 9.71 mA/cm^2, FF of 0.64, and η of 4.5%. Elbohy et al. developed graphene-embedded carbon nanofibers (GCNFs) by electrospinning a new class of counter electrodes, and GCNFs decorated with platinum nanoneedles (GCNFs-PtNNs) were prepared by redox reaction [42]. Overall device performance with mask of GCNFs was J_{sc} of 14.52 mA/cm^2, V_{oc} of 0.77 V, FF of 64%, and η of 7.15%. For GCNFs-PtNNs, the highest values were J_{sc} of 18.57 mA/cm^2, V_{oc} of 0.78 V, FF of 67%, and η of 9.70%. Saranya et al. studied the bimetallic (Fe–Ni) nanoparticle-embedded CNF-based counter electrode material for DSSCs [43]. The DSSC fabricated using Fe–Ni-embedded CNFs as counter electrode exhibited V_{oc} of 0.72 V, J_{sc} of 10.1 mA/cm^2, FF of 0.65, and PCE of 4.7%. Gupta et al. reported Cu-PACF/CNFs grown on activated carbon microfiber powder (PACF) and Cu NPs using CVD method [44]. The preparation of Cu-PACF/CNF is shown in Fig. 8.8a. The electrochemical impedance spectroscopy measurements revealed the high electrocatalytic activity and low charge transfer resistance (1.26 Ω cm^2). The fabricated device exhibited a power conversion efficiency of 4.36%.

Rameez et al. studied bimetal (Ni–Co) nanoparticles incorporated CNFs prepared via the electrospinning technique [45]. The photovoltaic performance values of assembled DSSCs were V_{oc} of 0.73 V, J_{sc} of 9.78 mA/cm^2, FF of 0.64, and PCE of 4.57%. Zhou et al. prepared graphene-beaded carbon nanofibers with incorporated Ni nanoparticles (G/CNFs-Ni) by the electrospinning technique [46]. The inclusion of graphene decreased the interfacial R_{ct} of 0.72 Ω cm^2. The G/CNFs-Ni composite exhibited a power conversion efficiency of 7.14%. Mohamed et al. reported CoCr-decorated carbon nanofibers prepared by a simple electrospinning technique [47]. The overall cell performance values were J_{sc} of 8.784 mA/cm^2, V_{oc} of 0.685 V, FF of 0.54%, and efficiency of 3.27%. Xiao et al. studied the Pt_3Ni alloy supported by carbon nanofibers (CNs) composites (Pt_3Ni/CNs) and Cr doped in Pt_3Ni alloy supported by carbon nanofibers (Cr-Pt_3Ni/CNs) composites both by using the solvothermal method [48], shown in Fig. 8.8a. From *J–V* results, the power conversion efficiency of Pt_3Ni/CNs is 8.34% and that of Cr-Pt_3Ni/CNs is 8.76% (Fig. 8.8b). Li et al. studied the surface-doped Pt_3Ni on carbon nanofibers (Pt_3Ni/CN) with molybdenum (Mo-Pt_3Ni/CN) as counter electrode in dye-sensitized solar cells [49]. The Mo-Pt_3Ni/CN exhibited the photovoltaic performance values of J_{sc} of 16.94 mA/cm^2, V_{oc} of 0.774 V, FF of 72.3%, and efficiency of 9.47% (Fig. 8.8c). Also, the lowest R_{ct} value of 2.04 Ω cm^2 indicates the highest catalytic activity due to an accelerated (I_3^-) reduction (Fig. 8.8d).

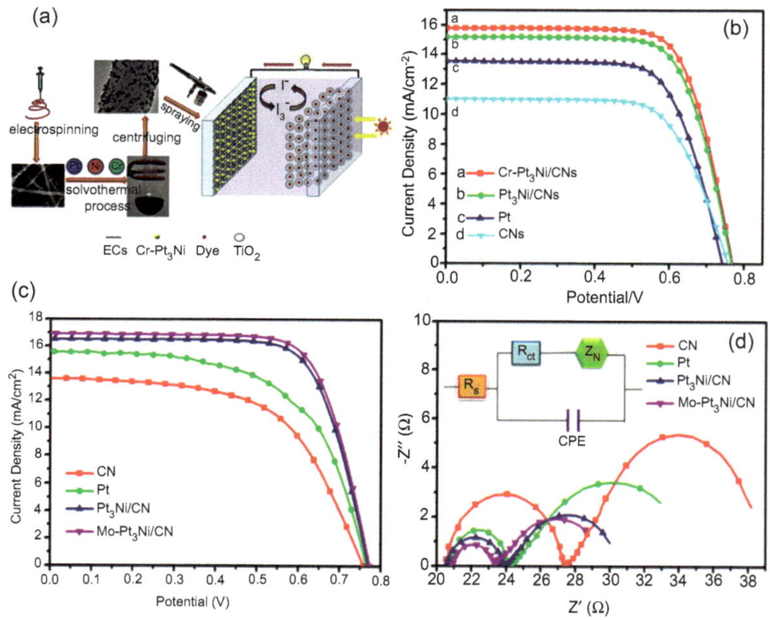

Figure 8.8 (a) Schematic representation of Cr-Pt_3Ni/CNs composite. (b) The *J–V* curves of DSSCs manufactured with Cr-Pt_3Ni/CNs composite. (c) The *J–V* curves of DSSCs manufactured with Mo-Pt_3Ni/CNs composite. (d) EIS curves of Mo-Pt_3Ni/CNs composite. Figures (a) and (b) reprinted from Ref. [48] and Figs. (c) and (d) reprinted from Ref. [49], Copyright 2016, with permission from Elsevier.

Table 8.2 Summary of photovoltaic performance of DSSCs fabricated using CNF-based composites as counter electrode

Counter electrode	J_{sc} (mA/cm^2)	V_{oc} (V)	FF	Efficiency (η, %)	Reference
ECN–Pt	4.35	0.83	67.0	8	[30]
Ni-CNT-CNF composite	15.83	0.80	63.0	7.96	[31]
Pd–Co-doped CNF	9.83	0.705	36.0	2.5	[32]
VACNF/Pt	15.18	0.67	71.0	7.22	[33]
CuNi NPs-doped carbon nanofibers	7.67	0.7	65.0	3.5	[34]
ECN:CNP (30:70)	13.37	0.79	62.8	6.64	[35]
Pt added ECN:TiO_2 (1:1)	13.47	0.84	66.0	7.46	[36]
WC/C-NF	15.35	0.76	67.0	7.77	[37]
N-doped TiO_2 nanoparticles within CNFs	15.65	0.67	60.50	6.31	[38]
ECNFs/PtNPs	13.33	0.83	68.8	7.6	[39]
NiCo NPs-doped carbon nanofibers	11.12	0.74	54.0	4.47	[40]
10 wt% CNFs-TiC	9.71	0.72	64.0	4.5	[41]
GCNFs-PtNNs	18.57	0.78	67.0	9.70	[42]
(Fe–Ni)-CNFs	10.1	0.72	65.0	4.7	[43]
Cu-PACF/CNF	11.12	0.75	54.0	4.36	[44]
(Ni–Co)–CNFs	9.78	0.73	64.0	4.57	[45]
G/CNFs-Ni	14.31	0.84	60.0	7.14	[46]
Co/Cr NPs-doped CNF	8.784	0.685	54.0	3.27	[47]

(*Continued*)

Table 8.2 (*Continued*)

Counter electrode	J_{sc} (mA/cm^2)	V_{oc} (V)	FF	Efficiency (η, %)	Reference
Cr-Pt_3Ni/CNs	15.80	0.770	72.0	8.76	[48]
Mo-Pt_3Ni/CN	16.94	0.774	72.3	9.47	[49]
Co-TiC NPs embedded on CNF	9.98	0.758	50.7	3.87	[51]
CoS/G-CNFs	10.25	0.72	66.0	4.9	[52]
Co_9S_8/CNFs	23.91	0.676	51.8	8.37	[53]
MoO_2/CNs	21.12	0.722	50.0	7.60	[54]

Kim et al. reported the synthesis of core/shell structured carbon nanofiber/platinum nanoparticle hybrid prepared via coaxial electrospinning [50]. The core/shell structure exhibited a power conversion efficiency of 4.6%. Yousef et al. studied the cobalt-titanium carbide nanoparticles (Co-TiC NPs) embedded on carbon nanofibers prepared by electrospinning [51]. The photovoltaic results showed that J_{sc}, V_{oc}, FF, and power conversion efficiency were 9.98 mA/cm^2, 0.758 V, 0.507, and 3.87%, respectively. Saranya et al. reported the growth of CoS nanoparticles onto graphitized carbon nanofibers (CoS/G-CNFs) by the combination of electrospinning and hydrothermal method as an efficient counter electrode for the DSSC [52]. The CoS/G-CNFs showed photovoltaic performance values of J_{sc}, V_{oc}, FF, and power conversion efficiency to be 10.25 mA/cm^2, 0.72 V, 0.66, and 4.9%, respectively. He et al. fabricated Co_9S_8 nanoparticles over the surface of CNFs via an electrospinning followed by hydrothermal reaction [53]. The assembled Co_9S_8/CNFs composite-based DSSC exhibited PCE of 8.37%. Li et al. reported the preparation of MoO_2/carbon nanofiber composites (MoO_2/CNs) by one-step hydrothermal method [54]. The schematic representation is shown in Fig. 8.9. The MoO_2/CN-based CE exhibited a photovoltaic conversion efficiency of 7.6%. These research output provides a better understanding and possible routes for reducing the usage of platinum as counter electrode without distressing the device efficiency.

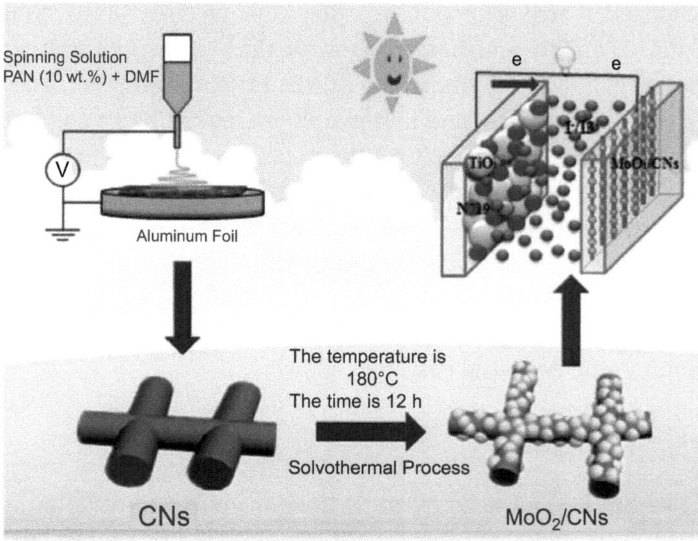

Figure 8.9 Schematic representation of MoO_2/CNs composite as counter electrode for dye-sensitized solar cells. Reprinted from Ref. [54], Copyright 2017, with permission from Elsevier.

8.6 Summary

Dye-sensitized solar cell is an efficient technology for the conversion of solar energy into electric energy with advantages of low cost, high efficiency, environment-friendly, and ease of fabrication. There are three key components of DSSCs, namely, photoanodes, electrolytes, and counter electrodes. The counter electrode is a significant component in influencing the overall device performance and cost. A counter electrode must possess some properties, such as good catalytic activity, high conductivity, and good stability. Platinum has these features, but it is expensive, and so researchers are in the pursuit of Pt-free counter electrodes for DSSCs. In that perspective, CNF is one of the members in the carbon family that exhibits an outstanding catalytic performance when applied as counter electrode for DSSC. CNFs are one-dimensional materials that contain well-organized graphitic layers in which the orientation is respect to the axis of the fiber this can be parallel, perpendicular, or stacked, which is known to be herringbone structure. Among the CEs based on CNF and their composites, the highest photoconversion efficiency of 9.70% is shown by GCNFs-PtNN-based DSSC. The ease of synthesis, low cost, and better stability make CNFs a good choice of material as CE for DSSCs. When compared to the Pt electrode, the main limitations of CNF-based counter electrode are its low conductivity and catalytic activity, which are still not better than those of Pt electrode. The future development must focus on the rational design and development of new composites that should satisfy the main requirements, such as conductivity, catalytic activity, stability, efficiency, and cost.

Acknowledgment

The author E. Vivek acknowledges the University Grants Commission (UGC), New Delhi for the Rajiv Gandhi National Fellowship (F1-17.1/2016-17/RGNF 2015-17-SC-TAM-13637).

References

1. O'Regan B and Grätzel M, A low-cost, high-efficiency solar cell based on dye-sensitized colloidal TiO_2 films, *Nature*, **353**, 1991, 737–740.
2. Hughes T V and Chambers C R, Manufacture of carbon filaments, U.S. Patent 405,480 (1889).
3. Robertson S D, Graphite formation from low temperature pyrolysis of methane over some transition metal surfaces, *Nature*, **221**, 1969, 1044–1046.
4. Baker S E, Colavita P E, Tse K-Y and Hamers R J, Functionalized vertically aligned carbon nanofibers as scaffolds for immobilization and electrochemical detection of redox-active proteins, *Chemistry of Materials*, **18**, 2006, 4415–4422.
5. Baker S E, Tse K-Y, Hindin E, Nichols B M, Clare T L and Hamers R J, Covalent functionalization for biomolecular recognition on vertically aligned carbon nanofibers, *Chemistry of Materials*, **17**, 2005, 4971–4978.
6. McKnight T E, Melechko A V, Austin D W, Sims T, Guillorn M A and Simpson M L, Microarrays of vertically-aligned carbon nanofiber electrodes in an open fluidic channel, *Journal of Physical Chemistry B*, **108**, 2004, 7115–7125.

7. McKnight T E, Melechko A V, Fletcher B L, Jones S W, Hensley D K, Peckys D B, Griffin G D, Simpson M L and Ericson M N, Resident neuroelectrochemical interfacing using carbon nanofiber arrays, *Journal of Physical Chemistry B*, **110**, 2006, 15317–15327.
8. Rodriguez N M, Chambers A and Baker R T K, Catalytic engineering of carbon nanostructures, *Langmuir*, **11**, 1995, 3862–3866.
9. Jong K P D and Geus J W, Carbon nanofibers: catalytic synthesis and applications, *Catalysis Reviews*, **42**, 2000, 481–510.
10. Tibbetts G G, Lengths of carbon fibers grown from iron catalyst particles in natural gas, *Journal of Crystal Growth*, **73**, 1985, 431–438.
11. Endo M, Kim Y, Hayashi T, Nishimura K, Matusita T, Miyashita K and Dresselhaus M, Vapor-grown carbon fibers (VGCFs), *Carbon*, **39**, 2001, 1287–1297.
12. Martin-Gullon I, Vera J, Conesa J A, González J L and Merino C, Differences between carbon nanofibers produced using Fe and Ni catalysts in a floating catalyst reactor, *Carbon*, **44**, 2006, 1572–1580.
13. Arshad S N, Naraghi M and Chasiotis I, Strong carbon nanofibers from electrospun polyacrylonitrile, *Carbon*, **49**, 2011, 1710–1719.
14. Zhou Z, Liu K, Lai C, Zhang L, Li J, Hou H, Reneker D H and Fong H, Graphitic carbon nanofibers developed from bundles of aligned electrospun polyacrylonitrile nanofibers containing phosphoric acid, *Polymer*, **51**, 2010, 2360–2367.
15. Lee K J, Shiratori N, Lee G H, Miyawaki J, Mochida I, Yoon S-H and Jang J, Activated carbon nanofiber produced from electrospun polyacrylonitrile nanofiber as a highly efficient formaldehyde adsorbent, *Carbon*, **48**, 2010, 4248–4255.
16. Zhou Z, Lai C, Zhang L, Qian Y, Hou H, Reneker D H and Fong H, Development of carbon nanofibers from aligned electrospun polyacrylonitrile nanofiber bundles and characterization of their microstructural, electrical, and mechanical properties, *Polymer*, **50**, 2009, 2999–3006.
17. Nataraj S, Kim B, Yun J, Lee D, Aminabhavi T and Yang K, Morphological characterization of electrospun carbon nanofiber mats of polyacrylonitrile containing heteropolyacids, *Synthetic Metals*, **159**, 2009, 1496–1504.
18. Trancik J E, Barton S C and Hone J, Transparent and catalytic carbon nanotube films, *Nano Letters*, **8**, 2008, 982–987.
19. Banks C E, Davies T J, Wildgoose G G and Compton R G, Electrocatalysis at graphite and carbon nanotube modified electrodes: edge-plane sites

and tube ends are the reactive sites, *Chemical Communications*, 2005, 829.

20. Banks C E, Crossley A, Salter C, Wilkins S J and Compton R G, Carbon nanotubes contain metal impurities which are responsible for the "electrocatalysis" seen at some nanotube-modified electrodes, *Angewandte Chemie International Edition*, **45**, 2006, 2533–2537.
21. Banks C E and Compton R G, New electrodes for old: from carbon nanotubes to edge plane pyrolytic graphite, *Analyst*, **131**, 2006, 15–21.
22. Shanmugam S and Gedanken A, Electrochemical properties of bamboo-shaped multiwalled carbon nanotubes generated by solid state pyrolysis, *Electrochemistry Communications*, **8**, 2006, 1099–1105.
23. Joshi P, Zhang L, Chen Q, Galipeau D, Fong H and Qiao Q, Electrospun carbon nanofibers as low-cost counter electrode for dye-sensitized solar cells, *ACS Applied Materials & Interfaces*, **23**, 2010, 572–577.
24. Veerappan G, Kwon W and Rhee S-W, Carbon-nanofiber counter electrodes for quasi-solid state dye-sensitized solar cells, *Journal of Power Sources*, **196**, 2011, 10798–10805.
25. Park S-H, Kim B-K and Lee W-J, Electrospun activated carbon nanofibers with hollow core/highly mesoporous shell structure as counter electrodes for dye-sensitized solar cells, *Journal of Power Sources*, **239**, 2013, 122–127.
26. Park S-H, Jung H-R and Lee W-J, Hollow activated carbon nanofibers prepared by electrospinning as counter electrodes for dye-sensitized solar cells, *Electrochimica Acta*, **102**, 2013, 423–428.
27. Chen C-S and Hsieh C-K, Oxygen-assisted low-pressure chemical vapor deposition for the low-temperature direct growth of graphitic nanofibers on fluorine-doped tin oxide glass as a counter electrode for dye-sensitized solar cell, *Japanese Journal of Applied Physics*, **53**, 2014, 11RE02.
28. Wang G, Kuang S and Zhang W, Helical carbon nanofiber as a low-cost counter electrode for dye-sensitized solar cells, *Materials Letters*, **174**, 2016, 14–16.
29. Zhao Y, Liu Y, Tong C, Ru J, Geng B, Ma Z, Liu H and Wang L, Flexible lignin-derived electrospun carbon nanofiber mats as a highly efficient and binder-free counter electrode for dye-sensitized solar cells, *Journal of Materials Science*, **53**, 2018, 7637–7647.

30. Poudel P, Zhang L, Joshi P, Venkatesan S, Fong H and Qiao Q, Enhanced performance in dye-sensitized solar cells via carbon nanofibers–platinum composite counter electrodes, *Nanoscale*, **4**, 2012, 4726.

31. Joshi P, Zhou Z, Poudel P, Thapa A, Wu X-F and Qiao Q, Nickel incorporated carbon nanotube/nanofiber composites as counter electrodes for dye-sensitized solar cells, *Nanoscale*, **4**, 2012, 5659.

32. Barakat N A, Akhtar M S, Yousef A, El-Newehy M and Kim H Y, Pd–Co-doped carbon nanofibers with photoactivity as effective counter electrodes for DSSCs, *Chemical Engineering Journal*, **211–212**, 2012, 9–15.

33. Mahpeykar S M, Tabatabaei M K, Ghafoori-Fard H, Habibiyan H and Koohsorkhi J, Low-temperature self-assembled vertically aligned carbon nanofibers as counter-electrode material for dye-sensitized solar cells, *Nanotechnology*, **24**, 2013, 435402.

34. Yousef A, Akhtar M S, Barakat N A, Motlak M, Yang O-B and Kim H Y, Effective NiCu NPs-doped carbon nanofibers as counter electrodes for dye-sensitized solar cells, *Electrochimica Acta*, **102**, 2013, 142–148.

35. Thapa A, Zhao Y, Poudel P, Elbohy H, Vaagensmith B, Zhang Z, Fong H and Qiao Q, Evaluation of counter electrodes composed by carbon nanofibers and nanoparticles in dye-sensitized solar cells, *IEEE Transactions on Electron Devices*, **60**, 2013, 3883–3887.

36. Sigdel S, Dubey A, Elbohy H, Aboagye A, Galipeau D, Zhang L, Fong H and Qiao Q, Dye-sensitized solar cells based on spray-coated carbon nanofiber/TiO_2 nanoparticle composite counter electrodes, *Journal of Materials Chemistry A*, **2**, 2014, 11448.

37. Jeong I, Lee J, Joseph K V, Lee H I, Kim J K, Yoon S and Lee J, Low-cost electrospun WC/C composite nanofiber as a powerful platinum-free counter electrode for dye-sensitized solar cell, *Nano Energy* **9**, 2014, 392–400.

38. An H-R, An H, Kim W-B and Ahn H-J, Nitrogen-doped TiO_2 nanoparticle-carbon nanofiber composites as a counter electrode for Pt-free dye-sensitized solar cells, *ECS Solid State Letters*, **3**, 2014, M33–M36.

39. Aboagye A, Elbohy H, Kelkar A D, Qiao Q, Zai J, Qian X and Zhang L, Electrospun carbon nanofibers with surface-attached platinum nanoparticles as cost-effective and efficient counter electrode for dye-sensitized solar cells, *Nano Energy*, **11**, 2015, 550–556.

40. Motlak M, Barakat N A, Akhtar M S, Hamza A, Kim B-S, Kim C S, Khalil K A and Almajid A A, High performance of NiCo nanoparticles-doped

carbon nanofibers as counter electrode for dye-sensitized solar cells, *Electrochimica Acta*, **160**, 2015, 1–6.

41. Saranya K, Subramania A, Sivasankar N and Mallick S, Electrospun TiC embedded CNFs as a low cost platinum-free counter electrode for dye-sensitized solar cell, *Materials Research Bulletin*, **75**, 2016, 83–90.
42. Elbohy H, Aboagye A, Sigdel S, Wang Q, Sayyad M H, Zhang L and Qiao Q, Graphene-embedded carbon nanofibers decorated with Pt nanoneedles for high efficiency dye-sensitized solar cells, *Journal of Materials Chemistry A*, **3**, 2015, 17721–17727.
43. Saranya K, Subramania A and Sivasankar N, Influence of earth-abundant bimetallic (Fe–Ni) nanoparticle-embedded CNFs as a low-cost counter electrode material for dye-sensitized solar cells, *RSC Advances*, **5**, 2015, 43611–43619.
44. Gupta R, Kumar R, Sharma A and Verma N, Novel Cu-carbon nanofiber composites for the counter electrodes of dye-sensitized solar cells, *International Journal of Energy Research*, **39**, 2014, 668–680.
45. Rameez M, Saranya K, Subramania A, Sivasankar N and Mallick S, Bimetal (Ni–Co) nanoparticles-incorporated electrospun carbon nanofibers as an alternative counter electrode for dye-sensitized solar cells, *Applied Physics A*, **122**, 2016, 122:71.
46. Zhou Z, Sigdel S, Gong J, Vaagensmith B, Elbohy H, Yang H, Krishnan S, Wu X-F and Qiao Q, Graphene-beaded carbon nanofibers with incorporated Ni nanoparticles as efficient counter-electrode for dye-sensitized solar cells, *Nano Energy*, **22**, 2016, 558–563.
47. Mohamed I M, Motlak M, Akhtar M S, Yasin A S, El-Newehy M H, Al-Deyab S S and Barakat N A, Synthesis, characterization and performance as a counter electrode for dye-sensitized solar cells of CoCr-decorated carbon nanofibers, *Ceramics International*, **42**, 2016, 146–153.
48. Xiao J, Cui M, Wang M, Sui H, Yang K, Li L, Zhang W, Li X, Fu G, Hagfeldt A and Zhang Y, Low-cost Cr doped Pt_3Ni alloy supported on carbon nanofibers composites counter electrode for efficient dye-sensitized solar cells, *Journal of Power Sources*, **328**, 2016, 543–550.
49. Li L, Wang M, Xiao J, Sui H, Zhang W, Li X, Yang K, Zhang Y and Wu M, Molybdenum-doped Pt_3Ni on carbon nanofibers as counter electrode for high-performance dye-sensitized solar cell, *Electrochimica Acta*, **219**, 2016, 350–355.
50. Kim G H, Park S H, Birajdar M S, Lee J and Hong S C, Core/shell structured carbon nanofiber/platinum nanoparticle hybrid web as a

counter electrode for dye-sensitized solar cell, *Journal of Industrial and Engineering Chemistry*, **52**, 2017, 211–217.

51. Yousef A, Brooks R M, El-Newehy M H, Al-Deyab S S and Kim H Y, Electrospun Co-TiC nanoparticles embedded on carbon nanofibers: active and chemically stable counter electrode for methanol fuel cells and dye-sensitized solar cells, *International Journal of Hydrogen Energy*, **42**, 2017, 10407–10415.
52. Saranya K, Subramania A, Sivasankar N and Bhargava P, In-situ growth of CoS nanoparticles onto electrospun graphitized carbon nanofibers as an efficient counter electrode for dye-sensitized solar cells, *Journal of Nanoscience and Nanotechnology*, **17**, 2017, 398–404.
53. Qiu J, He D, Zhao R, Sun B, Ji H, Zhang N, Li Y, Lu X and Wang C, Fabrication of highly dispersed ultrafine Co_9S_8 nanoparticles on carbon nanofibers as low-cost counter electrode for dye-sensitized solar cells, *Journal of Colloid and Interface Science*, **522**, 2018, 95–103.
54. Li L, Sui H, Zhao K, Zhang W, Li X, Liu S, Yang K, Wu M and Zhang Y, Preparation of carbon nanofibers supported MoO_2 composites electrode materials for application in dye-sensitized solar cells, *Electrochimica Acta*, **259**, 2018, 188–195.
55. Zheng Y, *Bioinspired Design of Materials Surfaces*, 2019, Elsevier.

Chapter 9

Quantum Dots as Emerging Counter Electrode Materials in Dye-Sensitized Solar Cells

A. Arivarasan
Department of Physics, Kalasalingam Academy of Research and Education, Krishnankoil, Virudhunagar, Tamil Nadu, India
arivarasan.nanotech@gmail.com

9.1 Dye-Sensitized Solar Cells

Dye-sensitized solar cells (DSSCs) have fueled up much attention recently due to their low manufacturing cost and high power conversion efficiency (PCE) [1–6]. A DSSC device consists of a dye-sensitized nanocrystalline TiO_2 photoanode, electrolyte containing triiodide/iodide (I_3^-/I^-) redox couple, and a counter electrode (CE) [7, 8]. The working mechanism of a dye-sensitized solar cell is illustrated in Fig. 9.1. The TiO_2 photoanodes have been extensively studied [9–11], while relatively less attention has been paid to CE, which is also critical for the DSSC device performance since its sluggish kinetics could cause serious charge recombination at the photoelectrode, resulting in a low PCE [12]. Platinum (Pt) is

Counter Electrode for Dye-Sensitized Solar Cells
Edited by Alagarsamy Pandikumar, Ramesh Mohan, and Kandasamy Jothivnekatachalam

ISBN 978-981-4877-38-1 (Hardcover), 978-1-003-11077-4 (eBook)
www.jennystanford.com

considered to be the best CE material because of its high catalytic activity toward reducing I_3^- ions [13–15]. However, Pt is a noble metal and the high cost limits its practical application. It is of great importance to develop low cost but well-performed CE.

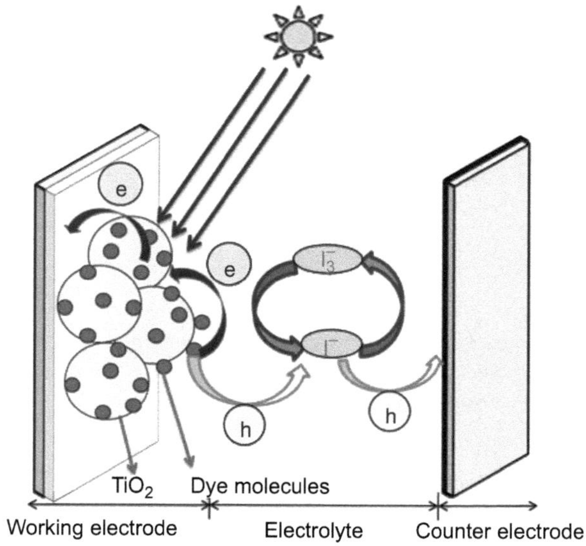

Figure 9.1 Working mechanism of dye-sensitized solar cells.

9.2 Counter Electrodes in Dye-Sensitized Solar Cells

An electrode is a solid electric conductor through which an electric current enters or leaves an electrolytic cell or other medium. An electrode in an electrochemical cell is referred to as either an anode or a cathode [16, 17]. The anode is now defined as the electrode at which electrons leave the cell and oxidation occurs (indicated by a minus symbol, "–"), and the cathode is defined as the electrode at which electrons enter the cell and reduction occurs (indicated by a plus symbol, "+"). For dye-sensitized solar cells, the anode is the electrode on which the metal oxide semiconductor is deposited and is also called photoanode, owing to the fact that incident sunlight often comes from here. The cathode is the electrode on which platinum and other conducting materials are deposited and is more often called counter electrode because it exists relative to the anode; a more important reason is that Gratzel first used "counter electrode" in his pioneering paper in *Nature* in 1991 [18].

In DSSCs, the CE undertakes the following three functions [19–23]:

(i) As a catalyst, it promotes the reduction of oxidized redox couple by accepting electrons at the surface of the CE and the reduction of oxidized dye by collecting electrons via ionic transport materials in solid state DSSCs.

(ii) As a positive electrode of primary cells, it collects electrons from the external circuit and transmits them into the cell. Thus, the ultimate function of the CE is to return the electrons from the external load back into the "circulation" within the cell.

(iii) As a mirror, it reflects the unabsorbed light from the cell back to the cell to enhance utilization of sunlight [24].

According to these basic functions, an optimal CE should possess the following qualities as required: high catalytic activity, high conductivity, high reflectivity, low cost, high surface area, porous nature, optimum thickness, chemical, electrochemical and mechanical stability, chemical corrosion resistance, energy level that matches the potential of the redox couple electrolyte, good adhesivity with TCO, and so on. Benchmark parameters for an ideal CE include 80% optical transparency at a wavelength of 550 nm, <20 Ω/sq sheet resistance (R_s), and 2–3 Ω cm^2 charge transfer resistance (R_{ct}) [25, 26]. The overvoltage should be as low as possible at photocurrent densities up to 20 mA/cm^2. However, it is not easy for one material to achieve all these parameters simultaneously in most of the cases. For example, the required R_{ct} and R_s can be achieved in carbon-based CEs, but optical transparency is not ideal.

9.3 Working Mechanism of Counter Electrodes

The CE has important effects on the photovoltaic parameter of DSSCs. The theoretical maximum photovoltage of the DSSC is determined by the energy difference between the redox potential of the mediator and the Fermi level of the metal oxide semiconductor on the photoanode and can be obtained only at zero current. Under load, however, the output voltage is usually less than the open circuit voltage. This voltage loss comes from the overall overpotential of the CE, which comes from the delivery of current through the electrolyte (the mass transfer overpotential) and through the electrolyte/counter electrode interface (the kinetic overpotential or charge transfer overpotential). The former is mainly influenced by the ionic conductivity of electrolytes and the transportation of mediator species from the CE to the photoanode, whereas the electrocatalytic properties of the CE surface toward mediator reduction will determine the magnitude of the latter [27, 28]. An effective CE should have good conductivity and exhibit high electrocatalytic activity for reduction of the redox couple. The reduction reaction ($I_3^- + 2e \rightarrow 3I^-$) on naked ITO (indium tin oxide) or FTO (fluorine-doped tin oxide) glass is extremely slow [23, 29]. Thus, in order to minimize the charge transfer overpotential, catalytic materials should be coated on conducting substrates to speed up the reaction.

In DSSCs, the transportation of charge carriers from the photoanode to the CE must meet various resistances, including [30–32]: the series resistance (R_s) comprising the sheet resistance of TCO glass and the contact resistance of the cell, the resistance at TCO/TiO_2 contact ($R_{TCO\text{-}TiO2}$), the transport resistance of electrons in the TiO_2 film (R_{TiO2}), the charge transfer resistance of the charge recombination between electrons in the TiO_2 film and I_3^- in the electrolyte (R_{ct}), the Warburg parameter describing the Nernst diffusion of I_3^- in the electrolyte (Z_d), the charge transfer resistance at the CE/electrolyte interface ($R_{CE\text{-}electrolyte}$), and the charge transfer resistance at the exposed TCO/electrolyte interface ($R_{TCO\text{-}electrolyte}$). In DSSCs, the charge transfer resistance at the CE/electrolyte interface ($R_{CE\text{-}electrolyte}$) is often dominant among multiple charge transfer resistances, and thus the R_{ct} often refers to $R_{CE\text{-}electrolyte}$, if there is no special note. Among these charge transfer resistances, R_s and the charge transfer resistance at the CE/electrolyte interface (R_{ct}) are enslaved to CEs. The series resistance of a solar cell dominates fill factor (FF) losses [33], especially in large area commercial solar cells. A smaller R_s will give a higher FF, which results in a high conversion efficiency [34–36]. The catalytic activity of the CE can be explained in terms of current density (J), which is calculated from the R_{ct}:

$$R_{ct} = RT/nFJ$$

Here, R, T, n, and F are the gas constant, temperature, the number of electrons transferred in the elementary electrode reaction ($n = 2$), and the Faraday's constant, respectively [37].

In order to improve the performance of DSSCs, many scientists have been devoting their research to CEs. Based on the above selection criteria, many materials have been chosen as the CE of dye-sensitized solar cells, and good progress has been achieved. Novel CE materials present excellent properties, even better than those of platinum, which are significant to promote the industrialization of dye-sensitized solar cells.

9.4 Preparation of Counter Electrodes

There are many kind of preparation methods for CEs in DSSCs, including thermal decomposition, electrochemical deposition, chemical reduction, chemical vapor deposition, hydrothermal reaction, sputter deposition, in situ polymerization, and so on. The preparation methods have a great influence on particle size, surface area, morphology, and the catalytic and electrochemical property of electrodes. The smaller particles and larger surface areas of electrodes will produce more catalytic active sites and promote the improvement of electrocatalytic activity of electrodes.

9.5 Requirements of Counter Electrode in DSSCs

The role of the counter electrode is to transfer electrons coming from the external circuit and to catalyze the reduction reaction of the oxidized electrolyte at the electrolyte/CE interface. If this reaction rate is unable to increase the rate of electron–hole pair generation, the collapse of the electrical equilibrium occurs in the form of accumulating holes in the internal cell. The accumulated holes might result in the chemical corrosion of photosensitizers (i.e., dye molecules) which significantly damages the stability of the cell. Therefore, the most important virtue of the CE is its ability to effectively catalyze the reduction of the electrolyte so that the reaction is facilitated as much as possible. To obtain superior catalytic properties, counter electrodes must have excellent electrical conductivity, high catalytic activity, and a large surface area. Conventionally, Pt-deposited FTO glass is the most frequently used CE in DSSCs, which usually works with triiodide as the electrolyte. Other types of CEs, such as metal sulfides, have also been studied to achieve high stability [38]. There were also some attempts to reduce the cost for manufacturing CEs because the conventional CEs consist of expensive noble metals, such as Pt and Au, and thus could be an obstacle to commercializing DSSCs in practical life. In this aspect, inexpensive candidates for CE materials have been reported, including metallic compounds [39], carbon derivatives [40], and even polymers [41].

Among the alternative cases of low-dimensional materials for photocatalytic applications, ultrasmall semiconductor nanocrystals (i.e., QDs) have recently emerged as a novel class of nanomaterials with unique physicochemical, electronic, magnetic, and optical properties. Owing to the higher surface area-to-volume ratio than their bulk counterparts, the recombination of the electron–hole pair within QDs drastically reduces as the particle size decreases. Therefore, QDs are expected to have higher photocatalytic activity than its bulk form [42]. Moreover, QDs get more and more attention because their confinement by the excited electrons and holes leads to optical and electronic properties different from those in bulk semiconductors. For QDs, the transportation length of electron/hole from crystal interface to the surface is short, which helps to accelerate the migration rate of electron/hole to the surface of QDs to participate in the reaction process [43].

9.6 Quantum Dots

Quantum dots are tiny particles, or "nanoparticles," of a semiconductor material with the size range from 0.1 to 10 nm (nanometers) in diameter (about the width of 1–100 atoms). Because of their small size, quantum dots display unique optical and electrical properties that are different in character to those of the corresponding bulk material. The most immediately apparent of these is the emission of photons under excitation, which are visible to the human eye as light. Moreover, the wavelength of these photon emissions depends not on the material from which the quantum dot is made, but its size. The size of QDs is tuned to emit photons in the visible region. The ability to control or tune the emission from the quantum dot by changing its core size is called the size quantization effect. The smaller the dot, it is closer to the blue end of the spectrum, and the larger the dot, it is closer to the red end. QDs can also be tuned to emit beyond visible light, that is, into the infrared or into the ultraviolet. At the end of the synthesis process, quantum dots appear physically either as a powder or in a solution.

9.6.1 Quantum Confinement Effect

A "quantum dot," defined for the first time at the beginning of 1980s by Ekimov [44], is a semiconductor crystal of nanoscopic size, whose excitons are confined in all three spatial dimensions. As a consequence of this confinement, its electronic structure can be tuned by varying the size of the crystal: the gap between the highest occupied molecular orbital (HOMO) and the lowest unoccupied molecular orbital (LUMO) widens with the decreasing crystallite size [45, 46] (phenomenon known as "size-dependant bandgap").

The size dependence of optical properties is determined by the degree of spatial confinement of the electronic wave function [47], which depends on relation between the size and the exciton Bohr radius. A NC of a given semiconductor material is considered a QD if its size is smaller than the exciton Bohr radius of the respective bulk semiconductor [48]. Under this size, it is possible to tune continuously the energy bandgap (E_g), therefore controlling the emission color or the absorption onset. A typical emission color of CdTe QDs with different sizes under visible and UV radiations is illustrated in Fig. 9.2. Another typical feature of QDs is the discrete structure of energy levels (termed "quantum size levels" – QSLs), replacing the continuous energy bands of a bulk material. In a spherical QD, surrounded by an infinite potential barrier, the energy of QSLs (characterized by angular momentum quantum number l) can be written in parabolic approximation as [49]:

$$E_{\mathrm{l,n}}^{\mathrm{e,h}} = \frac{h^2 \varnothing_{\mathrm{l,n}}^2}{2m_{\mathrm{e,h}} d^2}$$

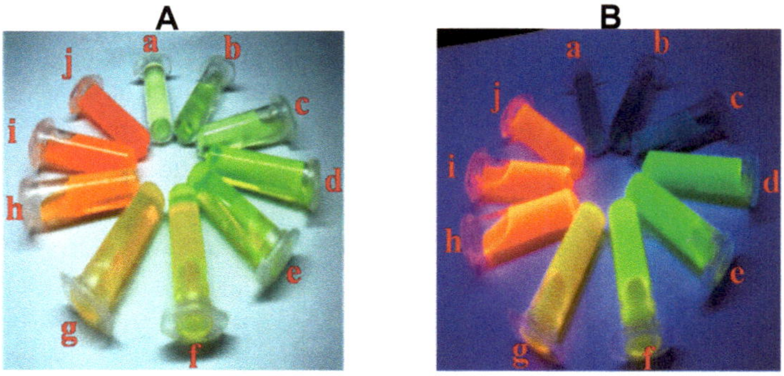

Figure 9.2 Size-dependent optical properties of CdTe QDs. (A) Under visible light. (B) Under UV radiation.

Here, $m_{e,h}$ are the electron and hole effective masses, d is the crystal size, and $\varnothing_{l,n}$ are the nth root of spherical Bessel function of order l.

9.6.2 Energy Levels

Quantum mechanics determines that only certain discrete energy levels are allowed within a single atom. If two identical atoms are held at large distances from each other, the electrons within each atom will have the same energy. As those two atoms are brought closer to one another, they interact, and no two electrons with the same spin can have the same energy. This governing principle is called the Pauli's exclusion principle. When a large number of atoms are brought together to form a solid, the discrete energy levels of the individual atoms are combine and becomes a continuous energy band. In bulk structures, these properties can be altered only by adding constituents to create defects, interstitials, or substitutions in the material, and it can be altered in nanostructures by tuning the size and shape of the materials. The energy level diagram of bulk materials with three kinds of nanostructured materials is shown in Fig. 9.3. In two-dimensional nanostructures, the step-like energy levels appear due to the electron confinement in one direction and the energy levels become atom-like discrete energy levels depending upon the electron confinement.

Quantum confinement is the phenomenon in which there is the widening of bandgap energy of the semiconductor material when its size is reduced to nanoscale. The bandgap of a material is the energy required to create an electron and hole pair with zero kinetic energy at a distance far enough apart that their Coulombic attraction could be ignored.

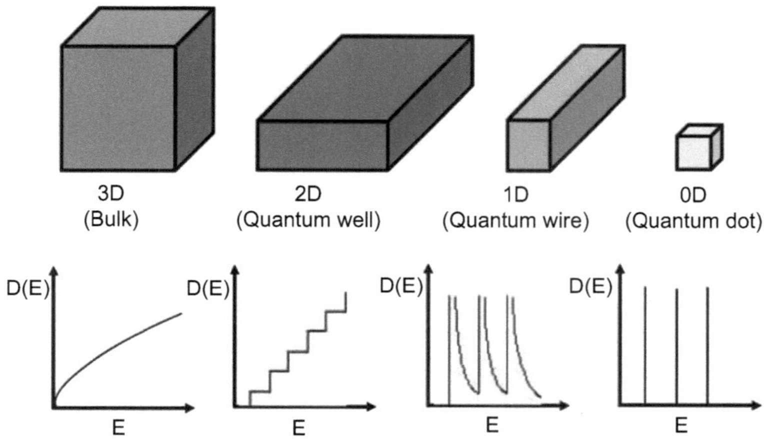

Figure 9.3 Energy level diagram of various nanostructured materials.

9.6.3 Emission Stokes Shift

Another distinctive feature of semiconductor NC optical properties is the redshift of the emission peak with respect to the absorption band edge, known as Stokes shift. The shift decreases with increasing the NC radius and disappears beyond a certain radius. According to Bagga et al. [50], the redshift is caused by the splitting of exciton states by the electron–hole exchange interaction: after absorption, the exciton in the singlet state first thermalizes to a triplet state, from where it de-excites to the original state, giving rise to a Stokes shift equal to the energy difference between singlet and triplet states.

9.6.4 Fluorescence Quantum Yield

In semiconductor nanocrystals, the high fluorescence quantum yield (defined as the ratio of photons emitted through fluorescence to photon absorbed) is achieved by maximizing the probability of band edge recombination; this is made difficult by the numerous competing processes that may occur after the generation of an exciton inside the nanocrystal:

- One of both charge carriers may be trapped in surface states [51]
- Scavenged by redox species [52] (if in solution)
- Become trapped on local defect sites [53] (if in a solid matrix)

In addition, electrons may undergo radiative or auger-assisted ionization, leading to non-radiative recombination or permanent oxidation of the nanocrystal [54]. Surface states originate from surface atoms not completely incorporated into the lattice, possessing orbitals that are capable of accepting or donating electrons. These orbitals (or dangling bonds) exhibit a range of energy levels that lie within the bandgap, providing non-radiative recombination pathways [55] and it competing with band edge luminescence. Passivation of these traps with appropriate surface ligands has been shown to greatly enhance the luminescence quantum yields of nanocrystals [56]; unfortunately, because of steric consideration, organic ligands have packing densities way lower than those of surface atoms [57]. In principle, formation of inorganic shell over the QDs completely eradicates surface traps and also reduces the probability of the other side processes that may compete with band edge radiative recombination.

9.7 Classification of Nanocrystals

❖ **Core Quantum Dots**

These are the simplest NCs, consisting of only one-phase semiconductor [58]. As a consequence of their size-dependent bandgap, the size of single-phase NCs can be related to their first exciton energy. In spherical NCs, the relation between radius r and bandgap $E_{\mathrm{G}}^{\mathrm{NC}}$ can be expressed using the formula from Brus et al. through an effective mass approximation:

$$E_{\mathrm{G}}^{\mathrm{NC}} = E_{\mathrm{G}}^{\mathrm{bulk}} + \frac{h^2}{2r^2} \cdot \left(\frac{1}{m_{\mathrm{e}}^*} + \frac{1}{m_{\mathrm{h}}^*} \right) - \frac{1}{r} \cdot \frac{1.8e^2}{\varepsilon}$$

Here, $E_{\mathrm{G}}^{\mathrm{bulk}}$ is the bandgap of the bulk material, m_{e}^* and m_{h}^* are electron and hole effective masses, respectively, h is the

Planck's constant, e is the elementary charge, and ε is the absolute permittivity of the bulk material.

❖ **Core/Shell of Core/Multi-Shell QDs**

Core/shell NCs (defined also as "nano-heterostructures") contain at least two semiconductor materials arranged in an onion-like structure [58]. Since optical properties of core/shell NCs depend on both core and shell sizes and composition, hypothetical curves that relate particle size (D) or ε with dimensions become multiple-variable functions and are generally not available in the literature. Growing an epitaxial shell provides a physical barrier between the inner "core" and the surrounding medium, making NCs less sensitive to surface chemistry and photooxidation. It also makes possible to tune their absorption/emission energies and their quantum yield [59]. The relative band edges of the two materials determine the properties of the resultant core–shell NCs; different combinations of core–shell materials result in nano-heterostructures that fall within two main categories [59]:

- Type-I structures: The bandgap of the shell material is larger than that of the core; both electrons and holes are confined in the core. Type-I nanocrystals are the ideal choice to achieve high quantum yield (QY), since they provide the greatest probability of band edge charge recombination.
- Type-II structures: Either the valence band edge or the conduction band edge of the shell material is located inside the bandgap of the core. Upon photoexcitation of the NC, this staggered band alignment leads to a spatial separation of charge carriers in different regions of the core–shell structure [59]; this reduces the overlap of the electron and hole wave functions and lengthens the lifetime of the excited state [60], thus increasing the probability that the unconfined charge carrier will be trapped or scavenged. Therefore, Type-II nanocrystals exhibit low quantum yields and they can be extremely sensitive to local environment.

Other categories can be defined as:

- Inverted Type-I [59]: The bandgap of the shell material is smaller than that of the core; depending on the thickness of the shell, charge carriers are partially or completely confined in the shell.

- Quasi Type-I [61]: One charge carrier is confined in the core (or in the shell) and the other is delocalized over the entire nano-heterostructure.

It is useful to point out that the relative band edges in a core–shell nanostructure change in function of both core and shell dimensions, since the alignment of quantized energy states is determined not only by bulk material energy offsets but also by confinement energies. Therefore, it can be possible to switch between Type-I and Type-II regimes also by varying the dimensions of the system [62, 63] in addition to its chemical composition.

9.7.1 Alloy Quantum Dots

The formation of an alloyed structure is an alternative way to influence the bandgap of NCs, changing their composition instead of their size. According to van Embden [51], alloying is simpler if anions of different semiconductors are the same, since this allows efficient diffusion of smaller cations between tetrahedral interstitial sites. A typical example is $Cd_{1-x}Zn_xSe$, corresponding to CdSe NCs, in which a fraction of Cd atoms is replaced by Zn ones in the crystal lattice. The bandgap of this ternary compound will be between those of pure CdSe and pure ZnSe [64, 65] according to Vegard's law:

$$E_{\mathrm{G}}^{\mathrm{Cd_xZn_{1-x}Se}} = xE_{\mathrm{G}}^{\mathrm{CdSe}} + (1-x)E_{\mathrm{G}}^{\mathrm{ZnSe}} - b(1-x)$$

Here, $E_{\mathrm{G}}^{\mathrm{CdSe}}$ and $E_{\mathrm{G}}^{\mathrm{ZnSe}}$ are the bandgaps of bulk CdSe and ZnSe, respectively.

The difference is not very pronounced in common anion systems [66, 67] (such as $Cd_{1-x}Zn_xSe$ or $Cd_{1-x}Zn_xS$), while it is more relevant in common cation compounds [68] (such as $CdSe_{1-x}S_x$ or $CdTe_{1-x}Se_x$), which consequently present larger bowing parameters.

9.8 Synthesis of Quantum Dots

There are two popular ways to synthesize quantum dots: one is top-down approach and the other is bottom-up approach. In the top-down approach, bulk semiconductor is thinned to form quantum dots. Several other processes have been involved in fabrication, such as electron beam lithography (EBL), reactive ion etching, focused ion beams, and dip pen lithography. The major shortcomings with these approaches include incorporation of impurities into the quantum dots and structural imperfections by patterning. Bottom-up approach means to synthesize the nanoscale materials by taking advantage of the chemistry and physics to artificially combine atoms and molecules in a nanoparticle cluster. In the bottom-up approach, QDs are fabricated in both physical and chemical methods.

9.8.1 Physical Methods

Physical methods for synthesizing quantum dots begin with steps in which layers are grown in an atom-by-atom process. For example, molecular beam epitaxy (MBE) has been used to fabricate consecutive layers and grow elemental, compound, or alloy semiconductor nanostructured materials on a heated substrate under ultrahigh vacuum ($\sim 10^{-10}$ Torr) conditions. Physical vapor deposition (PVD) grows layer by the condensation of solid from vapors produced by thermal evaporation or sputtering. Quantum dots can be self-assembled on a thin film by the chemical vapor deposition (CVD) technique.

9.8.2 Chemical Methods

By careful controlling of the parameters for a single solution or a mixture of solutions to precipitate, nuclei are generated and further nanoparticles growth may be achieved. Nucleation may be categorized as homogeneous or heterogeneous. Homogeneous nucleation occurs when solute atoms or molecules combine and reach a critical size without the assistance of a preexisting solid interface. By varying factors such as temperature, electrostatic double layer thickness, stabilizers or micelle formation, concentrations of precursors, ratios of anionic to cationic species, and solvent, quantum dot of the desired size, shape, and composition can be produced. Some of the common synthesis processes are the hydrothermal synthesis process, sol–gel process, microemulsion process, hot solution decomposition process, and microwave synthesis process.

9.8.3 Colloidal Synthesis

Colloidal semiconductor nanocrystals are synthesized from precursor compounds dissolved in solutions, much like traditional chemical processes. The synthesis of colloidal quantum dots is based on a three-component system composed of: precursors, organic surfactants, and solvents. When a reaction medium is heated to a sufficiently high temperature, precursors chemically transform into monomers.

Once monomers reach a high enough supersaturation level, the nanocrystal growth starts with a nucleation process. The temperature during the growth process is one of the critical factors in determining optimal conditions for the nanocrystal growth. It must be high enough to allow for rearrangement and annealing of atoms during the synthesis process while being low enough to promote crystal growth. Another critical factor that has to be stringently controlled during nanocrystal growth is the monomer concentration. The colloidal synthesis procedure is shown in Fig. 9.4.

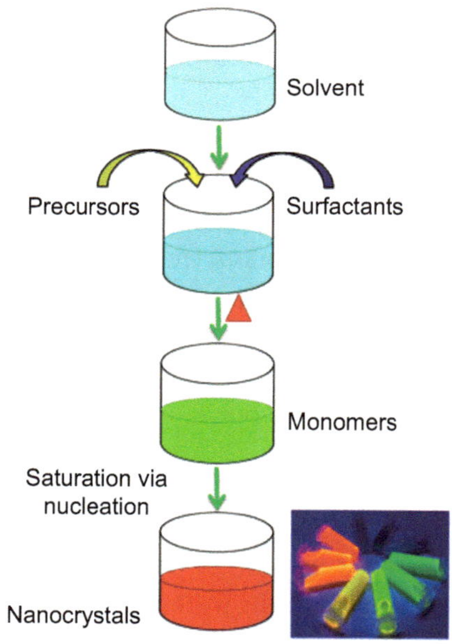

Figure 9.4 Schematic representation of colloidal synthesis procedures.

The growth process of nanocrystals can occur in two different regimes, that is, "focusing" and "defocusing." At high monomer concentrations, the critical size (the size where nanocrystals neither grow nor shrink) is relatively small, resulting in growth of nearly all particles. In this regime, smaller particles grow faster than larger ones (since larger crystals need more atoms to grow than smaller crystals), resulting in "focusing" of the size distribution to yield nearly monodisperse particles.

Quantum dots can be synthesized from various types of semiconductor materials (usually group II–VI: CdS, CdSe, CdTe, etc.; group III–V: InP, InAs, etc.; group IV–VI: PbSe, etc.) characterized by different bulk bandgap energies. The curves represent experimental data on the dependence of peak emission wavelength on QD diameter. The range of emission wavelength is 400–1350 nm, with size varying from 2 to 9.5 nm.

9.9 Properties of Quantum Dots

A unique property of nanomaterials is the high surface/interface-to-volume ratio. A high percentage of surface atoms lead to many size-dependent phenomena. The finite size of the particle confines the spatial distribution of electrons, leading to quantized energy levels due to the size effect. This quantum confinement has applications in various fields due to its special properties at the nanoscale.

9.9.1 Crystal Shape-Dependent Thermodynamic Properties

The large surface-to-volume ratio of nanocrystals greatly changes the role played by surface atoms in determining their thermodynamic properties. The increased surface atoms greatly increases the surface energy, which leads to the occurrences of atom diffusion at relatively lower temperature. The melting temperature of gold particles drops to as low as ≅300°C for particles with diameter smaller than 5 nm, which is much lower than the bulk melting temperature of about 1063°C for Au. The density of surface atoms changes significantly for different crystallographic planes, leading to different thermodynamic properties.

9.9.2 Magnetic Properties

The magnetic properties of nanoparticles differ from those of the bulk counterpart. The large surface-to-volume ratio results in a different local environment for surface atoms in their magnetic coupling/interaction with neighboring atoms, which leads to surface magnetic characteristics. Bulk ferromagnetic materials usually form with multiple magnetic domains and possess several small ferromagnetic particles consisting of only a single magnetic domain.

9.9.3 Mechanical Properties

Owing to the nanometer size, many of the mechanical properties of nanomaterials are enhanced to be different from those of the bulk counterpart. These properties include hardness, elastic modulus, fracture toughness, scratch resistance, fatigue strength, and so on. Enhanced mechanical properties of nanomaterials are due to the structural perfection. The small size renders a lower degree of internal structural imperfections, such as dislocations, micro-twins, and impurity precipitation.

9.9.4 Catalytic Properties

A larger percentage of surface atoms greatly increase surface activities. A unique surface structure, electronic states, and largely exposed surface area are required for stimulating and promoting chemical reactions. Gold (Au) nanoparticles possess excellent catalytic activity due to their high surface area. The size effect and the unusual properties of individual gold atoms determine the catalytic behavior of nanoparticles. Graphene also is a promising material for catalytic applications because it has a higher surface area of ~2580 m^2/g.

9.10 Preparation of QD Sensitizing Layer

QD layer is deposited on electrodes surface either through in situ method or through ex situ method.

9.10.1 In Situ Methods

In situ methods for QD sensitization over conducting surfaces are widely used for the fabrication of counter electrode for DSSCs. These are facile and low-cost techniques and also suitable for large-scale production. Two methods, namely, chemical bath deposition (CBD) and successive ionic layer adsorption and reaction (SILAR), are commonly used for in situ QD deposition.

- **Chemical bath deposition**

  Through the CBD technique, QD layer deposition can be completed in one bath, that is, container. Two, cationic and anionic, solutions are prepared separately. Then both the solutions are placed in one container, resulting into a bath solution for slow reaction of QD nucleation and growth. Transparent conducting electrode is then dipped into the bath solution for a specified period. Subsequently, in bath solution, QD nucleation and growth take place. QD deposition is controlled by varying dipping time, precursor solution composition, and reaction temperature. CBD facilitates the dense deposition of QDs on the electrode surface. As CBD is a time-consuming process, various research groups further modified CBD technique for fabrication of efficient and stable QDs as counter electrodes for DSSC. This sequential type CBD (SCBD) is quite similar to the SILAR technique [69]. In another approach, microwave-assisted CBD (MACBD) has been utilized for rapid nucleation and growth of QDs on electrode surface. Microwaves assist in quick and homogeneous heating to the aqueous solution, which leads to faster nucleation and growth of QDs, and for this the deposition time is optimized up to 20 min.

- **Successive ionic layer adsorption and reaction**

  SILAR technique is an extension of the CBD method. In SILAR, cationic and anionic precursor solutions are prepared separately and placed in two different containers. Transparent electrode is first dipped into cationic precursor, followed by rinsing and drying. Then, it is dipped into anionic precursor, again followed by rinsing and drying. The whole process of sequential dipping (of electrode into cationic and anionic precursor solutions each time followed by rinsing and drying) is referred to as one SILAR cycle. The desired QD growth depends upon the number of SILAR cycles, rate of dipping, dipping time, reaction temperature, and composition of precursor solutions. A comparative study of SILAR and CBD shows that SILAR has less processing time and maintains close stoichiometry formation [70]. Recently, a new type of SILAR termed as potential-induced ionic layer adsorption and reaction (PILAR) was evolved for in situ synthesizing and assembling of QDs into transparent electrode, and it enhanced the ion adsorption and assisted in high QD loading on electrode.

- **Electrochemical deposition**

  Using the electrochemical deposition (ECD) technique, thin layers of semiconductor QDs are grown on the substrate. The whole process is carried out in the presence of electric current. The substrate and the reference substrate are immersed into a bath solution consisting of salt precursor. On applying electric field, positive charge ions are attracted toward the negatively charged substrate. Therefore, QD layer is deposited on the substrate. The ECD method is a facile, low-cost technique for producing QD layer of desired thickness.

9.10.2 Ex Situ Methods

In ex situ methods, QDs are synthesized separately and further deposited into electrode surface. Molecular linker attachment (MLA), direct adsorption (DA), and electrophoresis are the commonly used ex situ techniques for QD deposition in DSSC.

- **Molecule linker attachment**

  In MLA, a bifunctional linker molecule such as 3-mercaptopropionic acid (3-MPA) is used for attachment of pre-synthesized QD to semiconductor surface. Carboxylic acid group present in the bifunctional linker molecule is attached to electrode surface, and at the other end of the bifunctional linker molecule, thiol group is interacted to QD. The technique enables the use of a selective size of pre-synthesized QD. In the pre-synthesis method of QD, it is prepared usually through hot injection method using a suitable capping agent (surfactant), such as mercaptopropionic acid (MPA), trioctylphosphine (TOP), trioctylphosphine oxide (TOPO), and oleic acid (OA). Capping agents assist in tailoring the structural, morphological, and optical characteristics of QD. Size and growth of QDs are controlled through concentration of the capping agent in precursor solution, reaction temperature, and reaction time.

9.10.3 Other Methods

QDs can also be deposited through other methods, such as DA, spray pyrolysis (SP), and CVD. DA is an ex situ technique and carried out by simply immersing transparent conducting electrode into colloidal QD solution for a specified time. QDs are deposited through DA without using any molecule linker or external DC voltage.

Each method discussed here for QD deposition has its own advantages and limitations. In situ fabrication, for example, CBD or SILAR, provides strong electronic coupling between QD and electrode, whereas ex situ methods, for example, EPD or MLA, enable to control size and shape of QD. However, the presence of linker molecules can hamper the effective charge transportation, leading to less efficient solar cell. Therefore, a proper deposition method can be selected depending upon the objective of the specified investigation.

9.11 Quantum Dots as Counter Electrodes

Quantum dots can be used as counter electrode materials due to their unique properties, such as wide absorption spectra with high absorption coefficient and narrow emission spectra. Owing to their high surface-to-volume ratio, they possess high surface activity and superior chemical stability. Recently, various types of nanomaterials in the form of quantum dots were utilized as the counter electrode materials for dye-sensitized and quantum dot-sensitized solar cells.

9.11.1 Metal Chalcogenide-Based Counter Electrodes

- **Copper chalcogenide counter electrodes**

  Copper sulfides have received enough attention as an important component in photovoltaic cells since the discovery of CdS/Cu_2S heterojunction solar cell in the work done by Reynolds et al. in 1954 [71]. Owing to the relatively high conductivity and catalytic activity to polysulfide electrolyte, copper sulfides are commonly used in QDSCs as CE materials. Cu_2S, a p-type semiconductor with bandgap of 1.1–1.4 eV, has been widely used as CE materials. Generally, Cu_2S CEs are fabricated by dipping the etched brass foils into polysulfide electrolyte to form compact Cu_2S film [72]. However, the poor stability and relatively low surface of the Cu_2S compact film still affect its performance as a CE. On account of those problems, novel Cu_2S CE preparation methods were developed in order to achieve CE with better stability, conductivity, and catalytic activity. Meng et al. applied the electroplating method to deposit a thin layer of copper on the surface of FTO and then immersed the copper coated glass in polysulfide solution to prepare Cu_2S CE [73, 74]. The favorable porous structure makes the film to have better catalytic ability. The interface charge transfer resistance of electrolyte and Cu_2S film is calculated to be a relatively low value of 5.7 Ω/cm^2. Based on the idea of increasing the surface area of copper sulfides to provide more active catalytic sites through the novel preparation method, copper sulfides with various morphologies were reported, including porous hierarchical structure [75], nanoplates [76], nanosheets [77], nanoparticles [78], and so on.

- **PbS counter electrodes**

  PbS is a well-known p-type semiconductor with a bandgap energy of about 0.4 eV in the bulk state. Owing to its light-harvesting and exciton-generation abilities over a wide range of light intensity, PbS has been used frequently as a photosensitizer in QDSSCs (quantum-dot-sensitized solar cells) [79, 80]. Some researchers, such as Lin et al. [81], have focused on its ability and utilized it as the CE material. The working mechanism of PbS CE-based DSSCs is shown in Fig. 9.5. Unlike the typical metal CEs used in DSSCs, the PbS CE generates a photocurrent by absorbing light as well as by transferring the delivered electrons to the electrolyte. This is quite meaningful because the light absorbed by the PbS CE is originally almost "wasted" light as this light penetrates the photoanode. To avoid the loss of open-circuit voltage (V_{oc}), photoanodes for DSSCs are usually sensitized by dye molecules with relatively large bandgap energy (e.g., Ru complex is the most widely used photosensitizer in DSSCs), which causes an inevitable waste of long-wavelength light. However, if semiconductors with relatively narrow bandgap energies are applied to CE, they can absorb the wasted light and generate an improved photocurrent. Furthermore, the addition of another Fermi level alignment of the p-type semiconductor at the CE/electrolyte interface results in the formation of another V_{oc}. This additional V_{oc} is combined with the original V_{oc} (with an energy level difference between the aligned Fermi levels in the photoanode and the electrolyte) to achieve a higher V_{oc}. Thus, by adopting the p-type PbS CE, Lin et al. achieved both a higher photocurrent and a higher open-circuit voltage at once [81].

PbS CE was deposited on FTO glass through the SILAR method by using precursor solutions of lead(II) nitrate and sodium sulfide aqueous solutions. By controlling the number of cycles in the SILAR method, the optimized thickness of deposited PbS was explored. As expected, the PbS CE exhibited the absorption of the entire range of visible light and even into the IR region, whereas the photoanode absorbed light only at a wavelength of about 700 nm. The superior electrocatalytic activity of the PbS CE was also proved by fitting the Nyquist plot from the EIS (electrochemical impedance spectroscopy) measurement and recording an R_{ct} value of about 50 W/cm^2, whereas the R_{ct} value of Pt was around 370 W/cm^2.

Owing to the improved working mechanism previously mentioned, the PCE of the DSSC containing the PbS CE was considerably increased. Compared with the Pt CE-based cell, CuS and PbS CE-based cells demonstrated increased values for J_{sc} (short-circuit photocurrent density) and V_{oc}.

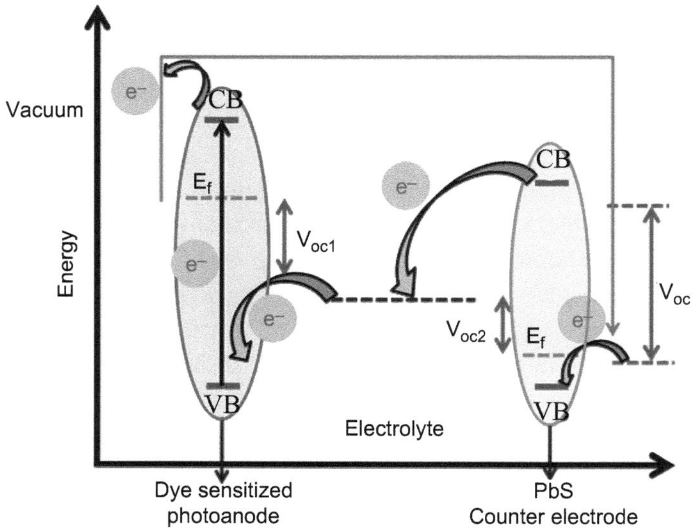

Figure 9.5 Working mechanism of PbS CE-based dye-sensitized solar cells.

❖ **Other metal chalcogenides**

Similar to Cu_2S and PbS QDs, various metal chalcogenides are employed as the active counter electrode materials in DSSCs. The term chalcogenide is more commonly reserved for sulfides, selenides, and tellurides, rather than oxides. Metal chalcogenides have various compositions, molecular structures, and distinctive properties, which make them a promising catalyst for the substitution of noble-metal-based electrocatalysts used in fuel cells, electrolysis of water, and DSSCs. An important application of chalcogenides in PV technologies is as optical absorption materials in QDSSCs [82–85]. Metal chalcogenides are also widely used as CEs in DSSCs [86–93].

Various metal sulfides, including CoS [87, 90, 94], MoS_2 [95], WS_2 [95], FeS_2 [96], and NiS_2 [97, 98], were reported as an alternative for Pt-based counter electrodes for DSSCs. With sulfide films as CEs, the DSSC achieved PCEs of 9.23% (CoS) and 9.65% (NiS). The values are obviously higher than that of DSSC with Pt CE (8.12%) under the same conditions. Similarly, various metal selenides are also reported as the counter electrode materials. The Tang's group systematically studied metal selenides (M–Se; M = Co, Ni, Cu, Fe, Ru) as CEs for bifacial DSSCs [99]. Owing to the superior charge transfer ability at the interface, good electrocatalytic activity toward triiodide reduction, and optical transparency, the bifacial DSSCs based on M-Se CEs yielded front and rear efficiencies of 7.64% and 5.05% for FeSe, 8.30% and 4.63% for $Co_{0.85}Se$, 7.85% and 4.37% for $Ni_{0.85}Se$, 6.43% and 4.24% for $Cu_{0.50}Se$, and 9.22% and 5.90% for $Ru_{0.33}Se$, respectively, in comparison with 6.18% and 3.56% of the device based on pristine Pt CE. Furthermore, they prepared and compared transparent $Ni_{0.6}Se$, $Ni_{0.7}Se$, and $Ni_{0.85}Se$ CEs. The DSSC with $Ni_{0.85}Se$ CE achieved impressive PCEs of 7.85%, 4.37%, and 10.63% for irradiation from front, rear, and both, respectively, which were superior to the cell performances from Pt CE [100]. Based on the $Ni_{0.85}Se$/mirror CE, Jia et al. achieved a high DSSC efficiency of 10.19% [101]. Various metal tellurides, such as CoTe [102] and $NiTe_2$ [103], are also used as the active counter electrode materials for DSSCs.

9.11.2 Graphene Quantum Dots

Owing to good catalytic activities toward I_3^- reduction, excellent resistances to iodine, high electric conductivities, earth abundances, and low costs, great progresses have been made in carbon-based CE materials. In particular, graphene is the challenging two-dimensional sheet of a single atomic layer of carbon with sp^2 hybridized orbitals, exhibiting excellent carrier mobility and conductivity, chemical inertness, and electrocatalytic activity [104]. As a new type of graphene nanostructure, graphene quantum dots (GQDs) exhibit good crystallinity and possess the same lattice structure as graphene [105]. Owing to quantum confinement and edge effects, GQDs have generated great interest both in the fundamental science and in potential applications in photovoltaic devices, fuel cells, and light-emitting diodes [106–109]. In particular, according to the recent research, the addition of GQDs could enhance the catalytic activity of CE, such as polyaniline/GQDs nanocomposite CE [110], GQDs doped polypyrrole (PPy) CE [111], and GQD-decorated mesoporous carbon aerogels composite CE [105].

The electrocatalytic activity of $SrRuO_3$ toward I_3^- reduction can be further improved by the incorporation of GQDs, in which

GQDs endowed with excellent intrinsic catalytic activity and high conductivity are incorporated into $SrRuO_3$ CE matrix by the dipping technique. Such an enhancement benefits from more active catalytic sites provided by GQDs and the formation of conductive network due to good connection between GQDs and $SrRuO_3$ nanoparticles. The predominant synergic effect of GQDs and $SrRuO_3$ CE deserves fast ion diffusions and collecting electrons from external circuit as well as electron transfer to catalytic sites. The champion device assembled with optimized SRO-GQD CE yields an enhanced PCE of 8.05%, much higher than that of the reference cell fabricated with Pt CE (7.44%). SRO-GQD CE also shows an excellent long-term electrochemical stability in the I_3^-/I^- electrolyte system [112].

Although GQDs have shown great promise as CEs for DSSCs, more attention is focused on hybrid-nanostructure-based CEs with GQDs to enhance the electrocatalytic activity. Li et al. [111] demonstrated a high-efficiency DSSC with GQDs doped into the PPy film as a hybrid CE. In this hybrid CE, it is shown that GQDs with nanometer size have more active defective sites at edges, which thereby exhibit an enhanced electrocatalytic activity and higher charge transfer rate. Meanwhile, the incorporation of GQDs results in a porous CE, which increases the number of electrocatalytic centres and enhances the I^-/I_3^- diffusion for better electrode kinetics. In summary, GQDs were explored to dope PPy via an electrochemical co-deposition method for fabricating GQDs doped PPy as a counter electrode for DSSCs. GQDs doped PPy film has more porous structure, more active sites, and higher charge transfer rate toward reduction of I_3^-/I^-. GQDs doping with 10% GQDs can greatly enhance the performance of DSSCs to achieve the highest PCE of 5.27%), which is ~20% higher than that of plain PPy- (4.46%) based DSSCs and is comparable to that of Pt CE-based device (6.02%). The GQDs doped PPy provides a promising candidate to replace Pt as an inexpensive counter electrode for high-performance DSSCs.

A novel nanoarchitecture consisting of GQDs@GFs as the CE for DSSCs was proposed by Chang et al. in 2016 [113]. In this designed nanostructure, zero-dimensional (0D) GQDs have high electrocatalytic activity and the high-quality GFs (graphene foams) act as the highly electrically conductive framework to enhance the charge transport between the 0D GQDs and the FTO substrate. A highly efficient hybrid GQDs@ graphene-foam-based CE consisting of 0D GQDs and GFs was fabricated. In the hybrid CE, the nanometer size 0D GQDs with more defective sites at the edges provided an outstanding electrochemical catalytic activity for the redox reaction of I_3^-/I^-; besides, GFs with extremely high electrical conductivity act as the fast electron transport channels for facilitating the charge injection between the 0D GQDs and the FTO substrate. This novel design strikes the appropriate balance between the electrical conductivity and the electrocatalytic properties of the graphene materials used for the CE in DSSCs. Compared with the traditional Pt electrocatalyst, the significant improvement of electrochemical properties is reflected in both the R_{ct} and the electrochemical stability under prolonged potential cycling. The DSSC with this hybrid CE exhibited an energy conversion efficiency (η) of 9.59% with V_{oc}, J_{sc}, and FF of 0.75 V, 19.23 mA/cm^2, and 66.50%, respectively.

9.11.3 Quantum Dots Sensitized Hybrid Counter Electrodes

In order to improve the performance and adaptability of CEs, much attention has been paid to developing hybrids or composite CEs. The hybrids are materials consisting of two or more components. The performance of hybrids could be improved by taking advantage of the synergetic effects that arise from different components of the hybrids. Currently, hybrids have become one of the popular CEs. As a result, various research studies have been reported on heterostructured CEs, such as Cu_2S/Pt [114], Cu_2S/RGO [115], CoS/graphene [116], PbS/graphene [117], Au/reduced graphene [118], $CuInS_2$/carbon composite [119], and MoX/CNT CEs [120].

Li et al. successfully prepared NiS_2@RGO nanocomposites via a facile in situ hydrothermal reaction and applied the composite as the novel CE material for DSSCs by a drop-casting technique [121]. The electrochemical results reveal that the positive synergetic effect between NiS_2 and RGO sheets on the reduction of I_3^- can greatly improve the catalytic activity, resulting in a significant enhancement of PCE. The PCE for the DSSC with NiS_2@RGO CE reaches 8.55%, which is significantly higher than that (7.02%) for the DSSC with the NiS_2 CE and that (3.14%) for the DSSC with the RGO CE. The nanocomposites are even better than the Pt electrode in solar cell performance, suggesting NiS_2@RGO is a promising alternative to Pt electrode for DSSCs.

A facile one-step hydrothermal method to prepare an efficient electrocatalyst from Cu_xS nanoparticles and RGO (Cu_xS@RGO) for polysulfide electrolyte reduction was developed by Hessein et al. in 2016 [122]. The Cu_xS nanocrystals that attached to RGO sheets were a mixture of CuS and $Cu_{1.8}S$ phases with high crystallinity, as clearly demonstrated by XRD measurement. The FESEM and Raman scattering measurements revealed a uniform distribution of Cu_xS nanoparticles along with their robust attachment on the graphene sheets. A durable and high-performance CE for CdS QDSSC applications was fabricated in a simple manner by direct coating of Cu_xS@RGO ink on an FTO substrate. The electrochemical measurements showed that the Cu_xS@RGO CE had superior electrocatalytic activity for polysulfide reduction to the Cu_2S/brass and Pt CEs. This is because of the large number of Cu_xS active sites and the rapid electron transport through the conductive graphene framework in the Cu_xS@RGO nanocomposite. As a result, the overall photovoltaic parameters of QDSSCs were greatly improved upon using our novel Cu_xS@RGO as a CE. A CdS QDSSC with a highly stable and reproducible PCE of 2.36% was obtained with the Cu_xS@RGO CE; this PCE is much higher than the values of 1.57% and 1.33% for Cu_2S/brass and Pt CEs, respectively. Thus, the developed Cu_xS@RGO CE has great potential for use in future QDSSC practical applications owing to its high productivity and excellent stability.

9.12 Characterization of Counter Electrodes

In order to understand the relative reaction process mechanism, such as charge generation, charge transfer, and charge recombination [123–126], to design novel materials and enhance the photovoltaic performance of DSSCs, a variety of characterization methods have been developed [127–129]. Electrochemical and photoelectrochemical methods have been considered as the most powerful tools in investigating the process mechanism, analyzing component interactions, and evaluating the performances of DSSCs [130, 131]. They include current–voltage (*I*–*V*) characterization, cyclic voltammetry (CV), EIS, intensity-modulated photocurrent spectroscopy (IMPS), and intensity-modulated photovoltage spectroscopy (IMVS).

9.13 Summary

The counter electrode is a key component and has a significant influence on both the photovoltaic performance and the device cost of DSSCs. As a counter electrode, it must possess high conductivity and good catalytic activity for electrolyte regeneration, as well as good stability. The platinum metal well satisfies all the requirements for CEs and has been the most frequently used CE material. However, the high cost and lack of stability limit its application widely. Among the Pt-free CEs, carbonaceous materials are the widest researched CE materials. The highest PCE (14.3%) of DSSCs is based on the FTO/Au/GNP (graphene nanoplatelet). The low cost, simple preparation, and good stability render carbon materials stronger competitors. The main drawbacks of the carbon-based CE are its relatively low conductivity and catalytic activity compared to the Pt electrode, the large dosage required to attain the targeted catalytic activity, and its poor adhesion to substrates. Conductive polymers are flexible, transparent, easily processed, and easily property tuned. Among conductive polymers, PEDOT exhibits the best performance, but its cost is high. The PPy-based CE is cheaper but the performance is inferior to that of PEDOT. The PANI-based CE has more attractive prospects thanks to its low cost and comparably better performance. The study of the TMC CE has begun later than other Pt-free CEs and it has become a hot research area in recent years owing to its material diversity, Pt-like catalytic activity, and easy preparation and modification. However, the performance of the TMC-based DSSC is relatively lower than that of the devices with other Pt-free CE materials, which might be explained by its poor conductivity for TMCs and electron transportation at interface of TMCs and substrates. Most hybrid-based DSSCs outperform the devices based on their corresponding components, which is attributed to the synergetic effects. The exact reason for the superior performance of the hybrids is not fully elaborated.

References

1. B. O'Regan and M. Grätael, *Nature*, 1991, 353, 737–740.
2. G. H. Guai, Y. Li, C. M. Ng, C. M. Li and M. B. Chan-Park, *Chem. Phys. Chem.*, 2012, 13, 2566–2572.
3. T. Chen, G. H. Guai, C. Gong, W. Hu, J. Zhu, H. Yang, Q. Yan and C. M. Li, *Energy Environ. Sci.*, 2012, 5, 6294–6298.
4. H. Pang, H. Yang, C. X. Guo, J. Lu and C. M. Li, *Chem. Commun.*, 2012, 48, 8832–8834.
5. A. Yella, H. W. Lee, H. N. Tsao, C. Y. Yi, A. K. Chandiran, M. K. Nazeeruddin, E. W. G. Diau, C. Y. Yeh, S. M. Zakeeruddin and M. Grätzel, *Science*, 2011, 334, 629–634.
6. G. Yue, J. Y. Lin, S. Y. Tai, Y. Xiao and J. Wua, *Electrochim. Acta*, 2012, 85, 162–168.
7. H. Yang, G. H. Guai, C. Guo, Q. Song, S. P. Jiang, Y. Wang, W. Zhang and C. M. Li, *J. Phys. Chem. B*, 2011, 115, 12209–12215.
8. K. S. Lee, J. Kwon, J. H. Im, C. R. Lee, N. G. Park and J. H. Park, *ACS Appl. Mater. Interfaces*, 2012, 4, 4164–4168.
9. A. Kay and M. Grätzel, *Chem. Mater.*, 2002, 14, 2930–2935.
10. W. Sharmoukh and N. K. Allam, *ACS Appl. Mater. Interfaces*, 2012, 4, 4413–4418.
11. Z. Yang, S. Gao, T. Li, F. Liu, Y. Ren and T. Xu, *ACS Appl. Mater. Interfaces*, 2012, 4, 4419–4427.
12. T. N. Murakami and M. Grätzel, *Inorg. Chim. Acta*, 2008, 361, 572–580.
13. V. Tjoa, J. Chua, S. S. Pramana, J. Wei, S. G. Mhaisalkar and N. Mathews, *ACS Appl. Mater. Interfaces*, 2012, 4, 3447–3452.
14. V. D. Dao, S. H. Kim, H. S. Choi, J. H. Kim, H. O. Park and J. K. Lee, *J. Phys. Chem. B*, 2011, 115, 25529–25534.
15. S. Y. Jang, Y. G. Kim, D. Y. Kim, H. G. Kim and S. M. Jo, *ACS Appl. Mater. Interfaces*, 2012, 4, 3500–3507.
16. S. Weinberg, *The Discovery of Subatomic Particles*, Revised Edition, Cambridge University Press, 2003, p. 81.
17. M. Faraday, *Philos. Trans. Royal Soc.*, 1834, 124, 77–122.
18. B. O'Regan and M. Gratzel, *Nature*, 1991, 353, 737–740.
19. K. Kalyanasundaram and M. Gratzel, *Coord. Chem. Rev.*, 1998, 77, 347–414.

20. S. Thomas, T. Deepak, G. Anjusree, T. Arun, S. Naira and A. Nair, *J. Mater. Chem. A*, 2014, 2, 4474–4490.

21. S. Yun, A. Hagfeldt and T. Ma, *Adv. Mater.*, 2014, 26, 6210–6237.

22. L. Wang, M. Al-Mamun, P. Liu, Y. Wang, H. Yang, H. Wang and H. Zhao, *NPG Asia Mater.*, 2015, 7, e226.

23. N. Papageorgiou, W. Maier and M. Gratzel, *J. Electrochem. Soc.*, 1997, 144, 876–884.

24. J. Wu, Y. Li, Q. Tang, G. Yue, J. Lin, M. Huang and L. Meng, *Sci. Rep.*, 2014, 4, 4028.

25. R. F. Service, *Science*, 2005, 309, 548–551.

26. J. Trancik, S. Barton and J. Hone, *Nano Lett.*, 2008, 8, 982–987.

27. N. Papageorgiou, *Coord. Chem. Rev.*, 2004, 248, 1421–1446.

28. K. Li, Z. Yu, Y. Luo, D. Li and Q. Meng, *J. Mater. Sci. Technol.*, 2007, 23, 577–582.

29. A. Hauch and A. Georg, *Electrochim. Acta*, 2001, 46, 3457–3466.

30. F. Fabregat-Santiago, J. Bisquert, E. Palomares, L. Otero, D. Kuang, S. Zakeeruddin and M. Gratzel, *J. Phys. Chem. C*, 2007, 111, 6550–6560.

31. B. Zhang, D. Wang, Y. Hou, S. Yang, X. Yang, J. Zhong, J. Liu, H. Wang, P. Hu, H. Zhao and H. Yang, *Sci. Rep.*, 2013, 3, 1836–1843.

32. N. Koide, A. Islam, Y. Chiba and L. Han, *J. Photochem. Photobiol. A*, 2006, 182, 296–305.

33. D. Pysch, A. Mette and S. Glunz, *Sol. Energy Mater. Sol. Cells*, 2007, 91, 1698–1706.

34. X. Yin, Z. Xue and B. Liu, *J. Power Sources*, 2011, 196, 2422–2426.

35. C. Chen, C. Chen and T. Wei, *Electrochim. Acta*, 2010, 55, 1687–1695.

36. S. Park, H. Boo and T. Chung, *Anal. Chim. Acta*, 2006, 556, 46–57.

37. T. Murakami and M. Gratzel, *Inorg. Chim. Acta*, 2008, 361, 572–580.

38. H. Yuan, J. Lu, X. Xu, D. Huang, W. Chen, Y. Shen and M. Wang, *J. Electrochem. Soc.*, 2013, 160, H624–H629.

39. Y. G. Tu, J. H. Wu, Z. Lan, Y. B. Lin, Q. Liu, B. C. Lin and G. Z. Liu, *J. Mater. Sci. Mater. Electron.*, 2014, 25, 3016–3022.

40. D. H. Youn, M. Seol, J. Y. Kim, J.-W. Jang, Y. Choi, K. Yong and J. S. Lee, *ChemSusChem*, 2013, 6, 261–267.

41. S. Abdul Al Mohsin, J. Armstrong and J. B. Cui, *J. Renew. Sustain. Energy*, 2012, 4, 043108.

42. H. R. Pouretedal, S. Narimany and M. H. Keshavarz, *Int. J. Mater. Res.*, 2010, 101, 1046.

43. D. Beydoun, R. Amal, G. Low and S. McEvoy, *J. Nanoparticle Res.*, 1999, 1, 439.
44. A. I. Ekimov and A. A. Onushchenko, *J. Exp. Theor. Phys.*, 1981, 34, 345–348.
45. D. V. Talapin, J.-S. Lee, M. V. Kovalenko and E. V. Shevchenko, *Chem. Rev.*, 2010, 110, 389–458.
46. C. B. Murray, D. J. Noms and M. G. Bawendi, *J. Am. Chem. Soc.*, 1993, 115, 8706–8715.
47. A. L. Efros and M. Rosen, *Annu. Rev. Mater. Sci.*, 2000, 30, 475–521.
48. V. I. Klimov, *Annu. Rev. Phys. Chem.*, 2007, 58, 635–673.
49. L. Brus, *J. Phys. Chem.*, 1986, 90, 2555–2560.
50. A. Bagga, P. Chattopadhyay and S. Ghosh, *Phys. Rev. B*, 2006, 74, 1–7.
51. J. van Embden, J. Jasieniak, D. E. Gomez, P. Mulvaney and M. Giersig, *Aust. J. Chem.*, 2007, 38, 457–471.
52. A. Issac, C. Von Borczyskowski and F. Cichos, *Phys. Rev. B*, 2005, 71, 161302.
53. R. M. Kraus, P. G. Lagoudakis, J. Müller, A. L. Rogach, J. M. Lupton, J. Feldmann, D. V. Talapin and H. Weller, *J. Phys. Chem. B*, 2005, 109, 18214–18217.
54. V. I. Klimov, *J. Phys. Chem. B*, 2000, 104, 6112–6123.
55. N. A. Hill and K. B. Whaley, *J. Chem. Phys.*, 1994, 100, 2831.
56. C. Bullen and P. Mulvaney, *Langmuir*, 2006, 22, 3007–3013.
57. J. E. B. Katari, V. L. Colvin and A. P. Alivisatos, *J. Phys. Chem. B*, 1994, 98, 4109–4117.
58. D. Dorfs, A. Eychmuller and A. L. Rogach, *Semiconductor Nanocrystal Quantum Dot*, Springer, Wien, New York, 2008, pp. 101–118.
59. P. Reiss, M. Protière and L. Li, *Small*, 2009, 5, 154–168.
60. S. Kim, B. Fisher, H.-J. Eisler and M. Bawendi, *J. Am. Chem. Soc.*, 2003, 125, 11466–11467.
61. C. deMello Donegá, *Chem. Soc. Rev.*, 2011, 40, 1512–1546.
62. D. Dorfs, T. Franzl, R. Osovsky, M. Brumer, E. Lifshitz, T. A. Klar and A. Eychmüller, *Small*, 2008, 4, 1148–1152.
63. L. P. Balet, S. Ivanov, A. Piryatinski, M. Achermann and V. I. Klimov, *Nano Lett.*, 2004, 4, 1485–1488.
64. J. E. Bernard and A. Zunger, *Phys. Rev. B*, 1986, 34, 5992–5995.
65. J. E. Bernard and A. Zunger, *Phys. Rev. B*, 1987, 36, 3199–3228.

66. M. Protière and P. Reiss, *Small*, 2007, 3, 399–403.
67. X. Zhong, Y. Feng, W. Knoll and M. Han, *J. Am. Chem. Soc.*, 2003, 125, 13559–13563.
68. R. E. Bailey and S. Nie, *J. Am. Chem. Soc.*, 2003, 125, 7100–7106.
69. S. Gorer and G. Hodes, *J. Phys. Chem. A*, 1994, 98(20), 5338–5346.
70. R. Vogel, K. Pohl and H. Weller, *Chem. Phys. Lett.*, 1990, 174(3–4), 241–246.
71. D. C. Reynolds, G. Leies, L. L. Antes and R. E. Marburger, *Phys. Rev.*, 1954, 96, 533–534.
72. T. Shen, J. J. Tian, L. L. Lv, C. B. Fei, Y. J. Wang, T. Pullerits and G. Z. Cao, *Electrochim. Acta*, 2016, 191, 62–69.
73. K. Meng, P. K. Surolia, O. Byrne and K. R. Thampi, *J. Power Sources*, 2014, 248, 218–223.
74. K. Meng, G. Chen and K. R. Thampi, *J. Mater. Chem. A*, 2015, 3, 23074–23089.
75. W. T. Liang, L. Q. Zhu, H. C. Liu, F. X. Xi and W. P. Li, *Electrochim. Acta*, 2015, 184, 285–294.
76. C. V. Thulasi-Varma, S. S. Rao, C. S. Kumar, C. V. Gopi, I. K. Durga, S. K. Kim, D. Punnoose and H. J. Kim, *Dalton Trans.*, 2015, 44, 19330–19343.
77. Y. Lei, H. M. Jia, Z. Zheng, Y. H. Gao, X. W. Chen and H. W. Hou, *CrystEngComm.*, 2011, 13, 6212.
78. S. M. Wang, W. W. Dong, X. D. Fang, S. Zhou, J. Z. Shao, Z. H. Deng, R. H. Tao, Q. L. Zhang, L. H. Hu and J. Zhu, *Electrochim. Acta*, 2015, 154, 47–53.
79. J. Jean, S. Chang, P. R. Brown, J. J. Cheng, P. H. Rekemeyer, M. G. Bawendi, S. Gradecˇak and V. Bulovic, *Adv. Mater.*, 2013, 25, 2790–2796.
80. N. Zhao, T. P. Osedach, L.-Y. Chang, S. M. Geyer, D. Wanger, M. T. Binda, A. C. Arango, M. G. Bawendi and V. Bulovic, *ACS Nano*, 2010, 4, 3743–3752.
81. C.-Y. Lin, C.-Y. Teng, T.-L. Li, Y.-L. Lee and H. Teng, *J. Mater. Chem. A*, 2013, 1, 1155–1162.
82. S. Ruhle, M. Shalom and A. Zaban, *ChemPhysChem*, 2010, 11, 2290–2304.
83. P. Kamat, *Acc. Chem. Res.*, 2012, 45, 1906–1915.
84. H. Jun, M. Careem and A. Arof, *Renew. Sustain. Energy Rev.*, 2013, 22, 148–167.
85. I. Hwang and K. Yong, *ChemElectroChem*, 2015, 2, 634–653.

86. J. Theerthagiri, A. Senthil, J. Madhavan and T. Maiyalagan, *ChemElectroChem*, 2015, 2, 928–945.

87. M. Wang, A. Anghel, B. Marsan, C. Ha, N. Pootrakulchote, S. Zakeeruddin and M. Gratzel, *J. Am. Chem. Soc.*, 2009, 131, 15976–15977.

88. J. Huo, J. Wu, M. Zheng and Z. Lan, *Electrochim. Acta*, 2016, 187, 210–217.

89. X. Xin, M. He, W. Han, J. Jung and Z. Lin, *Angew. Chem. Int. Ed.*, 2011, 50, 11739–11742.

90. C. Kung, H. Chen, C. Lin, K. Huang, R. Vittal and K. Ho, *ACS Nano*, 2012, 6, 7016–7025.

91. S. Peng, L. Li, H. Tan, R. Cai, W. Shi, C. Li, S. Mhaisalkar, M. Srinivasan, S. Ramakrishna and Q. Yan, *Adv. Funct. Mater.*, 2014, 24, 2155–2162.

92. Z. Jin, M. Zhang, M. Wang, C. Fen and Z. Wang, *Acc. Chem. Res.*, 2017, 50, 895–904.

93. J. Jia, J. Wu, J. Dong and J. Lin, *Electrochim. Acta*, 2015, 185, 184–189.

94. S. Chang, M. Lu, Y. Tung and H. Tuan, *ACS Nano*, 2013, 7, 9443–9451.

95. M. Wu, Y. Wang, X. Lin, N. Yu, L. Wang, A. Hagfeldt and T. Ma, *Phys. Chem. Chem. Phys.*, 2011, 13, 19298–19301.

96. Y. Wang, D. Wang, Y. Jiang, H. Chen, C. Chen, K. Ho, H. Chou and C. Chen, *Angew. Chem. Int. Ed.*, 2013, 52, 6694–6698.

97. H. Mulmudi, S. Batabyal, M. Rao, R. Prabhakar, N. Mathews, Y. Lam and S. Mhaisalkar, *Phys. Chem. Chem. Phys.*, 2011, 13, 19307–19309.

98. H. Sun, D. Qin, S. Huang, X. Guo, D. Li, Y. Luo and Q. Meng, *Energy Environ. Sci.*, 2011, 4, 2630–2637.

99. Y. Duan, Q. Tang, J. Liu, B. He and L. Yu, *Angew. Chem. Int. Ed.*, 2014, 53, 14569–14574.

100. Y. Duan, Q. Tang, B. He, R. Li and L. Yu, *Nanoscale*, 2014, 6, 12601–12608.

101. J. Jia, J. Wu, Y. Tu, J. Huo, M. Zheng and J. Lin, *J. Alloys Compd.*, 2015, 640, 29–33.

102. J. Guo, Y. Shi, Y. Chu and T. Ma, *Chem. Commun.*, 2013, 49, 10157–10159.

103. S. Patil, E. Kim, N. Shrestha, J. Chang, J. Lee and S. Han, *ACS Appl. Mater. Interfaces*, 2015, 7, 25914–25922.

104. K. S. Novoselov, A. K. Geim, S. V. Morozov, D. Jiang, Y. Zhang, S. V. Dubovos, I. V. Grigorieva and A. A. Firsov, *Science*, 2004, 306, 666–669.

105. C.-C. Wang and S.-Y. Lu, *Nanoscale*, 2015, 7, 1209–1215.

106. L. Li, G. Wu, G. Yang, J. Peng, J. Zhao and J. J. Zhu, *Nanoscale*, 2013, 5, 4015–4039.

107. L. Qian, Y. Zheng, J. Xue and P. H. Holloway, *Nat. Photonics*, 2011, 5, 543–548.

108. N. Maity, A. Kuila, S. Das, D. Mandal, A. Shit and A. K. Nandi, *J. Mater. Chem. A*, 2015, 3, 20736–20748.

109. X. Li, M. Rui, J. Song, Z. Shen and H. Zeng, *Adv. Funct. Mater.*, 2015, 25, 4929–4947.

110. M. Dinari, M. M. Momeni and M. Goudarzirad, *J. Mater. Sci.*, 2016, 51, 2964–2971.

111. L. Chen, C. X. Guo, Q. Zhang, Y. Lei, J. Xie, S. Ee, G Guai, Q. Song and C. M. Li, *ACS Appl. Mater. Interfaces*, 2013, 5, 2047–2052.

112. T. Liu, K. Yu, L. Gao, H. Chen, N. Wang, L. Hao, T. Li, H. He and Z. Guo, *J. Mater. Chem. A*, 2017, 5, 17848–17855.

113. Q. Chang, Z. Ma, J. Wang, P. Li, Y. Yan, W. Shi, Q. Chen, Y. Huang and L. Huang, *Energy Technol.*, 2016, 4, 256–262.

114. C. J. Raj, K. Prabakar, A. D. Savariraj and H.-J. Kim, *Electrochim. Acta*, 2013, 103, 231–236.

115. J. G. Radich, R. Dwyer and P. V. Kamat, *J. Phys. Chem. Lett.*, 2011, 2, 2453–2460.

116. H. Hu, J. Ding, J. Qian, Y. Li, L. Bai and N. Yuan, *Mater. Lett.*, 2014, 114, 7–10.

117. P. Parand, M. Samadpour, A. Esfandiar and A. I. Zad, *ACS Photonics*, 2014, 1, 323–330.

118. G. Zhu, L. Pan, H. Sun, X. Liu, T. Lv, T. Lu, J. Yang and Z. Sun, *ChemPhysChem* 2012, 13, 769–773.

119. X. Zhang, X. Huang, Y. Yang, S. Wang, Y. Gong, Y. Luo, D. Li and Q. Meng, *ACS Appl. Mater. Interfaces*, 2013, 5, 5954–5960.

120. M. Seol, D. H. Youn, J. Y. Kim, J.-W. Jang, M. Choi, J. S. Lee and K. Yong, *Adv. Energy Mater.*, 2014, 4, 1300775.

121. Z. Li, F. Gong, G. Zhou and Z.-S. Wang, *J. Phys. Chem. C*, 2013, 117, 6561–6566.

122. A. Hessein, F. Wang, H. Masai, K. Matsuda and A. A. El-Moneim, *Jpn. J. Appl. Phys.*, 2016, 55, 112301.

123. A. Hagfeldt, G. Boschloo, L. Sun, L. Kloo and H. Pettersson, *Chem. Rev.*, 2010, 110, 6595–6663.

124. J. Halme, P. Vahermaa, K. Miettunen and P. Lund, *Adv. Mater.*, 2010, 22, E210–E234.

125. A. Listorti, B. O'Regan and J. Durrant, *Chem. Mater.*, 2011, 23, 3381–3399.

126. P. Barnes, K. Miettunen, X. Li, A. Anderson, T. Bessho, M. Gratzel and B. O'Regan, *Adv. Mater.*, 2013, 25, 1881–1922.

127. S. Fantacci and F. De Angelis, *Coord. Chem. Rev.*, 2011, 255, 2704–2726.

128. A. Mishra, M. Fischer and P. Bauerle, *Angew. Chem. Int. Ed.*, 2009, 48, 2474–2499.

129. J. Ondersma and T. Hamann, *Coord. Chem. Rev.*, 2013, 257, 1533–1543.

130. H. Dunn and L. Peter, *J. Phys. Chem. C*, 2009, 113, 4726–4731.

131. W. Liu, L. Hu, Z. Huo and S. Dai, *Prog. Chem.*, 2009, 21, 1085–1093.

Index